THE STRANGE CASE

OF THE

RICKETY COSSACK

THE STRANGE CASE

OF THE

RICKETY COSSACK

AND OTHER CAUTIONARY TALES
FROM HUMAN EVOLUTION

IAN TATTERSALL

palgrave
macmillan

THE STRANGE CASE OF THE RICKETY COSSACK
Copyright © Ian Tattersall, 2015.
All rights reserved.

First published in 2015 by PALGRAVE MACMILLAN® TRADE in the
United States—a division of St. Martin's Press LLC, 175 Fifth Avenue, New
York, NY 10010.

Palgrave® and Macmillan® are registered trademarks in the United States,
the United Kingdom, Europe and other countries.

ISBN: 978-1-137-27889-0

Library of Congress Cataloging-in-Publication Data

Tattersall, Ian.
    The strange case of the Rickety Cossack : and other cautionary tales from
human evolution / Ian Tattersall.
        pages   cm
    ISBN 978-1-137-27889-0 (hardback)
        1. Paleoanthropology. 2. Human evolution. 3. Fossil hominids. I. Title.
GN281.T374   2015
599.93'8—dc23

                                                                        2014038032

Design by Letra Libre, Inc.

First edition: June 2015

10 9 8 7 6 5 4 3 2 1

Printed in the United States of America.

With gratitude to Elwyn Simons and Niles Eldredge

(both blameless).

CONTENTS

ACKNOWLEDGMENTS

THIS BOOK IS THE PRODUCT OF A LIFETIME'S INVOLVEMENT IN paleoanthropology, and of decades of interaction with numerous teachers, colleagues, and friends, often one and the same. It would be impossible to acknowledge everyone individually, so I will simply say thank you all, whether (or perhaps more likely not) you agree with everything I say here. I should also like to take this opportunity to acknowledge the many other people whose help was indispensable over the years, both as I carried out fieldwork in remote places and visited museum collections in urban centers. I have spent a lot of my career depending on the helpfulness of complete strangers, making me appreciate more fully the wonderful things of which our sometimes dubious species is capable. In terms of institutional support, my deepest appreciation must go to the American Museum of Natural History, which over the decades has provided an incomparable ambience in which to develop the viewpoints I have expressed here. At the American Museum, Jenn Steffey helped hugely in the preparation of the illustrations, and I would like also to acknowledge Don McGranaghan, Diana Salles, and the late Nick Amorosi, the other talented artists with whom I have had the pleasure of working at the AMNH over the years. John Van Couvering, Will Harcourt-Smith, and Jeanne Kelly kindly read the manuscript, and all were civil enough to limit their comments to the constructive. At Palgrave Macmillan this project started at the initiative of Luba Ostashevsky, and was nurtured to fruition by Elisabeth Dyssegaard, with whom it has been a great pleasure to work. I am also grateful to Bill Warhop for his sensitive copyediting and to Donna Cherry and Lauren LoPinto for shepherding the book through the production process.

PREFACE

I AM APPROACHING THE END OF A LONG CAREER AS A PALEO-
anthropologist—a student of human evolution—with a nagging feeling of
incompletion. Not that I would trade that career for anything: it has been
hugely fulfilling and has provided me with an endless supply of incident—
however unwanted on occasion—that has guaranteed the absence of bore-
dom. Even more important, over the last half century profound changes
in the field of paleoanthropology have made it a wonderfully exhilarat-
ing place to be. Most obvious among these are the huge additions made to
the human fossil record, the traditional archive of our evolutionary past,
nowadays closely rivaled by the advent of powerful molecular genetic techniques that allow us alternative ways of glimpsing our biological history.
But numerous other new technologies have also given us unprecedented
approaches to clarifying the ages and the lifeways of our various hominid
precursors, and novel viewpoints on how evolution works have offered us
new ways of thinking about our biological record.

Yet for all the excitement, in some respects paleoanthropology has remained curiously static compared to other areas of paleontology. Indeed,
it seems fair to say that today's paleoanthropologists are in general a lot
more like their precursors of the mid-twentieth century than, say, modern dinosaur or fish paleontologists are. Perhaps this is inevitable, since
it is particularly difficult to escape from preconception when looking at
our own egocentric species, *Homo sapiens,* and its extinct relatives. What is
more, we tend to scrutinize the evidence for our own past in considerably
more detail than we do that of other species. Still, for all the extenuating
circumstances, the science of human evolution has not borne the burden of
history lightly; in this field, more than most, what we think today continues

to be very intensely influenced by what we thought yesterday—and the day before that.

I realized this fact with particular force a few years ago while writing *Masters of the Planet,* a book in which I attempted to put together a coherent account of just how, from a remote starting point as a bipedal but otherwise unremarkable ape, we human beings contrived to become, rather rapidly, the extraordinary creatures we are. As the writing progressed I realized that if I was to provide a narrative of human evolution that would make ready sense, I would have to omit any substantial mention of the convoluted—and in many ways highly insular—histories of discovery and ideas in paleoanthropology. This was a serious omission since, given the sheer weight of paleoanthropology's historical burden, it left a huge gap in the story. It is that gap that led to the book you are holding now, which is in effect a complement to the earlier one. It is an idiosyncratic history of paleoanthropology that is intended to show just how received wisdoms about human evolution have always conditioned what we have believed about our own origins, often in the face of compelling evidence to the contrary. For the half century or so during which I have actively participated in the field, I have tried to make this point by describing the development of my own ideas about evolutionary processes and about the fossil record they have produced, while enlivening the narrative with some anecdotal experience. My hope is that by the book's end I will have convinced you that how we interpret the process by which we became human really does matter, because it so greatly impacts our view of ourselves.

Perhaps you will allow me to begin with what may initially appear to be a digression.

LEMURS AND THE DELIGHTS OF FIELDWORK

THE SAIL FIRST APPEARED AS A TINY TRIANGLE ABOVE THE SHIM-
mering horizon. Gradually it resolved into the image of a rickety dhow,
butting slowly nearer through the ocean swells. At first, though, I barely
noticed it. I had other preoccupations. Sitting forlornly on a gleaming deserted beach on Mohéli, the smallest of the Comoro Islands, which sit in the
middle of the vast Mozambique Channel between Madagascar and Africa, I
was actually wondering if I would ever see the outside world again.

Three weeks earlier, I had arrived in Moroni, capital of Grande Comore Island, after one of the more hair-raising aircraft rides of my life. I
had shown up early in the morning at Dar es Salaam airport, in neighboring Tanzania, for the scheduled short flight aboard one of Air Comores'
two ancient DC-4s. Once at the airport I was told that Air Comores hadn't
actually been seen for a month, but that if I felt optimistic I could wait. So
I waited, along with two unkempt French youths, one of whom sported a
bushy golden beard and long, lank hair that made him resemble that weird
image on the Shroud of Turin. Hours of desultory conversation later, a battered DC-4 with a star and crescent on its tail finally rolled up to the ramp
and spluttered to a stop in the blazing midday sun. For what seemed an
eternity, nothing happened. Finally, the flight to Moroni was called, and we

three optimists duly trooped out to the aircraft. As we climbed the rolling steps its door opened, emitting a blast of hot, fetid air that almost knocked us off our feet. Inside, we found a cabin crammed with heat-prostrated passengers, and our way forward blocked by oil drums lashed down along the aisle. Not an empty seat in sight. We looked around at the cabin steward. He nodded at the barrels.

Once we were settled uncomfortably on our improvised seats, the door closed and the aircraft taxied out. After what felt like an eternity, it staggered into the air. Circling up to a cruising altitude of what seemed no more than a couple thousand feet, we headed out over the choppy Mozambique Channel, every whitecap sharply visible below us. Arriving in Moroni, a bone-jarring landing took up the entire runway, and after we reached the terminal it took me awhile to disentangle myself from all the junk in the aircraft interior. By the time I finally got inside the tiny building, the Air Comores captain was already seated at the bar at the end of the narrow room, fiercely grasping a tumbler brimming with neat whisky.

Recognizing him from earlier times—Air Comores flight crews in those days were grizzled veterans who had flown Dien Bien Phu, Biafra, and Katanga, had seen it all, and usually liked to reminisce—I greeted him warmly. He didn't respond in kind. Instead, he raised his eyes from his glass, fixed me with a wide-eyed stare, and said, "We were two tons over maximum gross weight on takeoff, and if one engine had so much as hiccoughed, we'd all be dead!" Taken aback, I said, "But, Monsieur X, why do you fly under these conditions?" His answer: "Monsieur, I am 75 years old. Who else will pay me to fly?"

That was the Comoros in a nutshell. Assembled one by one into the French Empire at various dates between 1840 and 1912, the four tiny islands of the archipelago (Grande Comore, Mohéli, Anjouan, and Mayotte) had earlier been independent sultanates lying at the far end of the Arab trading route down the east African coast from Oman. During colonial times they were administered as a dependency of the huge but equally remote French island of Madagascar, making them about as forgotten and neglected as it's possible to be. But isolation and tranquility are not the same thing; and once Madagascar had achieved its own independence in 1960, the Comoros embarked on a tortuous history, the complexities of which have been far out of proportion to the archipelago's size.

Until 1974 the islands remained under the wing of distant France, but at the end of that year a referendum on independence was held. Three islands voted to go it alone; but the southernmost island of Mayotte, which had been French the longest, voted to stay that way. Ultimately this was allowed, and in response the enraged Comorian authorities blockaded the wayward island. I was living in Mayotte at the time, and I clearly remember my growing dismay as first the beer, and then the cigarettes, and then all civilized forms of food ran out. Eventually, we were reduced to subsisting off a supremely tedious ratatouille of obscure local vegetables. When, after some months, we finally learned that a shipment of imported food had beaten the embargo and arrived in one of the island's few primitive stores, we were less than entirely amused to discover that the entire consignment consisted of cans of ratatouille niçoise. Times have changed, though; since 2011 Mayotte has been an Overseas Department of France, complete with supermarkets, paved roads, high prices, a bloated bureaucracy, and a polluted lagoon.

In the other Comoros, in contrast, time has in many ways stood pretty still—even if events haven't. As if to make up for a marked lack of economic progress, though, politics have been very lively indeed. Almost as soon as the archipelago had unilaterally declared its independence, its president was deposed in a coup—the first of four—by the notorious French mercenary Bob Denard, already a veteran of mischief in Algeria, Katanga, Rhodesia, and other trouble spots. Following a short period of confusion, a revolutionary named Ali Soilih was installed as president, and things got really interesting. Ali was a francophobe Maoist with peculiar ideas about democracy. To purge the Comoros of its colonial heritage he supervised the burning of all the government's civil status records; and to assure his legitimacy he reduced the voting age to 14 (he was a Che Guevara–like idol to kids)—a maneuver that allowed him to squeak by in a referendum on his rule.

It was in the midst of all of this folderol that I found myself sitting despondently on that beach in Mohéli. One rarely hears anything about the Comoros in the outside world, and when I'd arrived in Moroni I had been blissfully unaware of all the chaos and mayhem going on. All I had known was that the islands were unusually hard to get to in that summer of 1977. And this is where my little fieldwork débâcle becomes relevant to what I shall be writing about in the rest of this book. I had come to the

Comoros to continue the studies of lemurs I had begun some years earlier in Madagascar. Lemurs, as you may know, are the primates unique to that huge island; and they are of particular interest because they last shared a common ancestor with humans around 50 million years ago. That's a pretty long time, and while the lemurs are nonetheless our cousins, it makes them remoter from us than monkeys and apes are. At the same time, while the lemurs have diversified enormously in their island redoubt, in some important respects they have remained a lot more primitive than humans have. This means that if you want to know how your own ancestors lived and functioned early in the Age of Mammals, it is to the lemurs you have to turn. And if—like me—unkind fate had excluded you from Madagascar (which is another story), you were out of luck. With one exception. Possibly because they were transported there by humans within the last thousand years or so, a couple of species of lemurs (out of well over 50) have managed to make it over to the Comoros and have established wild populations there. As a result, a lemurologist without access to Madagascar had little option but to head for the archipelago.

So there I was. But the instant I stepped off the airplane in Moroni I knew I had made a huge mistake. There was a creepy feeling to the place, and the air was full of little bits of paper ash—the result, I soon learned, of burning all those property, birth, and marriage records. What's more, it proved hard to find any of the bureaucrats from whom I had been used to seeking permits and logistical support. So, since I knew that there were no lemurs on Grande Comore, it was clear that I should head with all dispatch to another of the islands where there were lemurs to be found, and where in my previous experience nothing much ever happened anyway.

So much for experience. Off I went to the local Air Comores office and bought a one-way ticket to Mohéli, the next island in the chain. There were no formalities on departure, but stepping off the aircraft onto the dirt strip at my destination, I found myself greeted by a couple of teenagers, one of whom was toying with a Kalashnikov. The other gave me a broad smile, introduced himself as a member of the Revolutionary Youth, and invited me to stay in the organization's hotel, at an exorbitant rate. Hotel? That was something new for Mohéli. Still, not for the first or the last time in my life I had been made an offer I couldn't wisely refuse, and I joined my hosts in a military jeep driven by Kalashnikov. After a bumpy few minutes we arrived at our destination, which turned out to be the ruins of what not long before

A female mongoose lemur from Mohéli, in the Comoro Islands. Eulemur mongoz belongs to a genus that contains at least five other living species, and by some reckonings as many as eleven. This makes it a pretty typical mammal, and reminds us that it is very unusual to be the sole living member of one's genus, as we are. Drawn by Nick Amorosi.

had been the well-appointed home of an ylang-ylang plantation manager. Now it was a disaster zone. The plumbing had been ripped out, the generator and wiring had been stolen, and the roof was sagging. I was taken to a filthy room devoid of anything except rats, rubbish, and piles of old bricks, and was invited to pay a week's rent in advance. With my cash in their pockets my hosts bid me farewell, and vanished in their jeep.

At this point, my state of mind was not the best. I wandered out of my hovel and walked several dusty kilometers into Fomboni, the island's main and only town. It was almost deserted, but I was able to purchase a couple of cans of Spam and some rice and matches in a ramshackle store before I returned to my grand residence. That night, Baco Mari turned up. Baco was a young man who had driven me around on his official motorcycle during my previous residence on Mohéli, and who was now full of deeply unsettling revolutionary fervor. Disappointingly, he informed me that the mongoose lemurs I had previously studied had disappeared from the neighborhood—as, with massive immigration of refugees, they later did from almost the entire island. Gone was any reason to remain in Mohéli.

The next day thus found me at the Air Comores office, asking for a ticket to the next island, Anjouan. The agent amiably said he was willing to sell me one, but that he couldn't do so without seeing my exit visa. Exit visa? I was traveling within the same country. And nobody had asked me for an exit visa when I had left Moroni. "Nonetheless, Monsieur," I was assured, "you need an exit visa to leave Mohéli. We cannot issue you a ticket without one." And where would I get this visa? Why, from the Revolutionary Youth, of course.

The same evening, well after dark, I received a visit from an emissary of the dignified and courtly gentleman who had been Préfet of Mohéli before independence. In hushed tones my visitor invited me to visit the Préfet at once, but warned I should not be seen for fear of reprisals on both sides. I duly sneaked over to the Préfet's darkened house, where I found the former official in a state of extreme distress. He explained to me that the government had been disbanded, and that the island was now being terrorized by the submachine gun–wielding teenagers of the Revolutionary Youth, Ali's version of Mao's Red Guards. Their authority was backed up by a unit of the Tanzanian military that lurked in a camp a little way along the coast. Nothing happened in Mohéli that the Revolutionary Youth didn't want, the Préfet told me; and a lot of bad things were happening. "Monsieur," he said, "you have to leave as soon as possible!"

But to comply, I had to have that exit visa. And that meant facing the Revolutionary Youth. I trudged down to Fomboni again and found a gaggle of 14-year-olds of both sexes chatting, smoking pot (which Ali Soilih had legalized), and idly fondling an assortment of evil-looking weapons. I requested an exit visa, and was immediately refused. When I asked why, the teens replied that they didn't have a rubber stamp. Logical enough, I suppose, and on my numerous subsequent visits the answer was invariably the same.

So I had pretty good reasons for my despondency as that dhow hove into view. It was obvious that the Youth weren't prepared to issue me my exit visa as long as they had a cash cow occupying their ruin, and I had no idea what they would do when my money was gone. Nor was I anxious to find out: it might get really ugly if I couldn't pay my rent. What's more, with no access to lemurs I had absolutely nothing to distract me from my plight, especially after I had run out of batteries for my shortwave radio. Naturally enough, then, that dhow on the horizon became an object of particular interest. Gradually it grew bigger, until it became clear that it was heading for the beach more or less right where I was. When it ground into the sand a few yards away from me, it disgorged a pair of very seasick figures whom I instantly recognized as my young French companions from the flight from Dar es Salaam to Moroni. They duly joined me in my slum, and next day we trooped down together to the Revolutionary Youth's headquarters.

Once he had recovered from his seasickness, the Jesus figure, he of the golden beard and tumbling locks, turned out to be a charismatic character who knew, intuitively, just how to charm the Youth. Within a few days he had them eating out of his hand, and one of the kids eventually obtained a piece of linoleum and carefully carved an exit stamp into it. Our passports duly decorated, the three of us took the next flight to Anjouan—and eventually out to the wider world.

Anyone who knows the lovely Comoro Islands and their good-natured people—a blend of African, Arab, and Malagasy influences with a dash of errant European added—must wish them a more settled and more prosperous future. But meanwhile, the outlook for the archipelago's unusual flora and fauna remains dire. When I first went to the Comoros in 1974, the total human population of all four islands in the archipelago was reckoned to be about 250,000; today just the three islands of the Union of the Comoros are home to three times that many people, maybe more. Even in Mayotte, where the environment is threatened by development rather than by poverty, there has been a marked drop in lemur numbers over the past decade. This matters, because most lemur populations are under huge and increasing pressure in Madagascar. Mayotte is their only homeland controlled by a developed nation that—if it wished—could easily afford to protect them.

Yet for all the neglect that has been lavished on them, the lemurs must occupy a very special place in the heart and mind of anyone interested in the long evolutionary history of humankind. This is because, as I've hinted, in certain respects they resemble our own remote ancestors of the Eocene epoch, around 50 million years ago. Ranging today from animals the size of a mouse to that of a large house cat (and, until not very long ago, up to the size of a gorilla), lemurs have smaller brains relative to their body sizes than we "higher" primates do, and they depend much more on the sense of smell. Still, they are clearly not as dumb as these attributes have led many to suppose. Primate psychologists have tended to evaluate lemur intelligence using tests developed for monkeys and even people: forms with much greater manual skills, and a strong tendency to evaluate objects entirely visually rather than also by smell. This has been a huge drawback to understanding how exactly lemurs apprehend the world; and as researchers devise ways of testing the lemurs' cognitive skills in ways more appropriate to their own ways of relating to the environment around them, we can hope to gain valuable insight into the kind of cognition that preceded our own unusual way of dealing with information. In this context alone, the lemurs have an enormous amount to teach us, as they do about our ancient ancestors' social and ecological strategies. But in my own particular case, they had an even more significant lesson to impart.

Most of my research nowadays is in paleoanthropology, as I try to understand the fossil and archaeological evidence for human evolution. Like all of my contemporaries in the English-speaking world, I was initially trained to look upon the biological history of the human family as a single-minded (and implicitly heroic) struggle from primitiveness to perfection. The going assumption, when I was a graduate student, was that evolution is a process of fine-tuning that, over the eons, gradually makes its subjects ever more perfectly adapted to the environments in which they live. That perspective was hardly surprising, since it intuitively appeals to members of a species that is the only representative of its group in the world today. This lonely state of ours makes it appear logical to reconstruct the story of our evolution by projecting the single species *Homo sapiens* back in time, in a single, gradually modifying lineage. And of course, in one very limited sense this view is accurate; for we are certainly the product of a unique series of ancestors, each of which existed over a definable period of past time. But that is purely in hindsight; in prospect, which is how evolution works, things would have looked very different. The upshot is that what I'd been taught about human evolution was very far indeed from the whole story, and it was my involvement with lemurs, beginning in the late 1960s, that gave me my very first inkling of this.

What any observer of the lemurs of Madagascar immediately notices, before anything else, is that they are amazingly diverse. There are way north of 50 species of these lovely primates, arrayed into five different families. These range from the tiny scurrying mouse lemurs, to the cat-sized quadrupedal "true" lemurs, to the long-legged leaping sifakas and lepilemurs that typically hold their bodies vertical, to the bizarre, bat-eared, bushy-tailed aye-aye. What's more, if you'd been fortunate enough to visit Madagascar a mere 2,000 years ago, you would additionally have seen the hanging "sloth lemurs," the giant koala-like megaladapids, and the vaguely monkey-like archaeolemurids, all of them weird and wonderful, and much bigger than their surviving relatives.

In other words, the whole lemur fauna loudly blares diversity at you. Not just diversity in body forms and lifestyles, but evolutionary diversity too, with several distinctive families, lots of genera, and huge numbers of species. What's more, it turns out that diversity, even on this eye-catching scale, is hardly unusual among successful groups of mammals. In fact, it is rather routine. Successful mammal families have a strong tendency to spread geographically and to diversify phylogenetically. Applying this lesson from my studies of the lemurs to the very varied human fossil record, I soon realized that our own hominid family is actually no exception. Far from having been the linear, perfecting process that most of us were taught about—if we were taught anything about it at all—human evolution actually witnessed high drama, as one new species after another was pitchforked out into the ecological arena to do battle for survival and success—and, as likely as not, to become extinct in the process.

Acknowledging this pattern of events entirely changes our perspective on how we became the highly unusual creatures we are. For it rapidly becomes apparent that we human beings are not the burnished product of incremental improvement over the eons. Instead, we are one particular outcome of an active process of experimentation with the evidently many ways there are to be a hominid. This, in turn, casts considerable doubt on the received assumption that we were fine-tuned by evolution to be a creature of a particular kind.

Yet because there is undeniably something very unusual about us today—so unusual, indeed, that our species really *has* radically changed its relationship to Nature—it has proven very difficult for paleoanthropologists to perceive our hominid precursors as just another bunch of primates, and to see the process that produced us as merely another example of something that occurs widely among mammals. Right from the point at which it was realized—over a century and a half ago—that we have a fossil record, there has been a distinct leaning toward what one might call hominid exceptionalism: the instinctive assumption that, because we are so different today, our own ancestors did not necessarily play the evolutionary game by the same rules that apply to everything else out there. It is this vague exceptionalist feeling that accounts for the notable conservatism in paleoanthropology that I began by complaining about. In one form or another, it has existed from paleoanthropology's very earliest beginnings—where any account of our understanding of ourselves has to start.

CHAPTER 1

HUMANKIND'S PLACE IN NATURE

WE HUMAN BEINGS HAVE ALWAYS TAKEN WHO WE ARE PRETTY much for granted. After all, we are so unlike the other living creatures with which we share the world that, on the face of it, there's very little room for confusion. Bizarrely, we walk upright on two legs, freeing up hands that have acquired astonishing powers to manipulate things. We lie to each other using that highly unusual medium we call language. We have vast brains, housed in skulls that are precariously balanced atop rather puny bodies; and we are not at all discomfited by our ability to entertain conflicting beliefs within those brains. We wield amazingly complex technologies, without which we couldn't possibly any longer get by, and we process information in a totally unprecedented fashion. The list of our unique features is endless: nothing else in nature looks or does business anything like us, with the result that for most of its history the idiosyncratic *Homo sapiens* has hardly seen any need to define itself. The differences between us and the other animals around us seem so obvious that, for example, back in the eighteenth century the great lexicographer Samuel Johnson thought it adequate to define a "man" as a "human being," and "human" as "having the qualities of a man." Perhaps this was not the high point of Johnson's magisterial *Dictionary;* but at a time when our closest known relative in

Nature was the poorly understood orangutan, Johnson's contemporaries hardly needed more.

Certainly Johnson's equally famous contemporary Karl Linnaeus didn't. Linnaeus, the father of modern zoological classification, is revered for having taken the bold and revolutionary move of classifying us among the primates, along with the lemurs and the monkeys and the apes. But when it came to actually describing *Homo sapiens,* he abandoned his normal practice of listing features that would help his readers to recognize a member of a particular species if they saw one. Instead, he contented himself with the admonition *nosce te ipsum:* "Know thyself." And perhaps this vagueness was entirely forgivable. After all, while we are dimly aware that we are integrated into the natural world, it is equally evident that, in some complex and occasionally unfathomable ways, we are also set apart from it. We just don't function in quite the same way as other animals do, and for all that we are fundamentally just another mammal species, with hearts and kidneys and gallbladders and a need to eat and breathe, there is undeniably something *different* about us.

The first savant we know of who tried to figure out just how our strange species fits into the world was Aristotle, back in fourth-century-BC Greece. As the earliest comparative anatomist on record, Aristotle was interested in the continuities he perceived among all living forms. He even saw continuity between the animate and inanimate worlds, and was the first to propose that life had been formed from an inert precursor such as pond sludge (which, as shown by the much later invention of the microscope, actually teems with life). Still, the universe Aristotle envisaged was fixed and eternal, its every component fixed in a hierarchy from the simplest to the most complex. On the lowest rungs of the "ladder of being" were rocks and other objects that possessed no life force, but just existed. Above these lay the simplest animate things like plants, with the properties of life, growth, and reproduction. Yet higher up the ladder lay the various animals, with additional capacities such as movement and memory. Highest of all were humans, with the power of reason. Every kind of organism had its particular position on the ladder, and the whole sequence was somehow initiated by a "Prime Mover" whose exact qualities remained ambiguous.

Closer to our own times, Aristotle's arrangement was eagerly seized upon by the Scholastic theologians who dominated medieval Christian thought. With Saint Augustine, these scholars were only too happy to equate Aristotle's prime mover with the biblical God, who presided over a Great Chain of Being in which every living thing occupied its divinely preordained place. Human beings ranked below the various kinds of angels, but above the lions and other savage beasts that lorded it over the meeker domestic animals, and so on, down the line. Just as a physical chain is a continuum that nevertheless consists of discrete links, the Chain of Being connected human beings to the rest of God's creation while also holding them apart from it: a clever ambiguity that helped explain the "rudely wise and darkly great" human condition so admiringly mocked by Alexander Pope in his 1734 *Essay on Man,* still the most penetrating portrait of the human predicament ever penned.

Nonetheless, as early as the sixteenth and seventeenth centuries, naturalists were already trying to refine the Scholastics' ritualized description of the living and inanimate worlds. In the eighteenth century it was Linnaeus's great genius to recognize not only that the living world was clearly structured, but that this structure could best be classified using an arrangement of sets-within-sets. This insight allowed Linnaeus to give us the system of classifying living things we still use, in which our species, *Homo sapiens,* belongs to a genus *Homo* that in turn belongs to the family Hominidae, which forms part of the order Mammalia, and on up until we are united with all Animalia. In this inclusive hierarchy, each category includes everything below it in the scale, so that *H. sapiens* is only one of several species—the others are now extinct—that are classified in the genus *Homo,* while the family Hominidae contains several genera, and so on. This is a significant difference from the military-style hierarchy represented by the Great Chain of Being, in which every species occupies only a single rank. The inclusive Linnaean hierarchy turned out to capture very nicely the pattern of historical events that we now know gave rise to the Tree of Life to which we belong.

In Linnaeus' time, as today, the basic unit into which living things were seen to be "packaged" was the species. As early as the seventeenth century, the English naturalist John Ray had recognized that what gives any species its (occasionally permeable) borders is that it is bound together as a reproductive unit. In today's parlance we would say that, among sexually reproducing organisms, the species is the largest population within which interbreeding can freely take place. Of course, in most cases species are also physically distinct in some way from their close relatives; but because recognizable varieties often exist *within* species that will readily interbreed if and when they get the chance, the key test of species membership is reproductive continuity—whether the members of a group are interested in mating with each other, and can do so successfully. In other words, individuals don't belong to the same species because they look similar; they look similar because they belong to the same species.

Since we human beings are an integral part of the living world, appreciating just where and how we fit into the biota requires not only that we know ourselves through introspection, but that we understand exactly what species are, how we can recognize them, and how they may or may not change over time. This is because, whatever your organism of interest, if you don't first have a reliable family tree that properly links it to its closest relatives and on to the rest of the biota, it is tough to say anything about where it came from. The same thing goes for your expectations about how change has occurred in evolutionary time, because your model of change has to fit the actual facts of history. And, because what we think we know scientifically about the world may change, we forget at our peril that received wisdom—what we were taught—requires continual reexamination.

The one thing the Scholastics had in common with Ray and Linnaeus was the notion of a stable, unchanging universe, in which each species had its own immutable place. But by the time the early years of the nineteenth century came around, some scientists were beginning to have their doubts about this. The toils of early geologists and paleontologists, in particular, were beginning to raise questions about the fixity of both landscapes and fossil species. Sedimentary rocks pile up on each other like layers in a cake, but while you can, with luck, figure out the local succession of strata fairly straightforwardly, correlating them from one place to another can be very tricky because the physical composition of a rock is no guarantee of its age. Faced with this reality, early geologists did not take long to hit on the expedient of using the fossils contained in sedimentary rocks (usually the bones and teeth of vertebrates, or the shells of mollusks and other marine invertebrates) to determine the order in which they were deposited. This is possible because different faunas characterize different periods of geological time—as we now know, because of evolutionary change.

Still, even before ideas of evolution became current, it was already clear not only that the planet Earth has had a very long history, but that it did not always look as it does today. One early way of accounting for the differences observed between ancient and modern faunas without too greatly violating religious belief was to view fossil species as the victims of ancient "catastro-phes." Analogous to the biblical Flood, such catastrophes were hypothetical events that had carried off entire earlier creations. And in the scant but rapidly accumulating fossil record, there were indeed hints of large-scale faunal replacements. These came in the form of evidence for the very occa-sional mass-extinction events that saw the sudden disappearance of a large proportion of life forms on Earth. By and large, the biblical faithful (who at least nominally included almost everyone in Europe) found such readings an acceptable fudge; but by the early nineteenth century, new perspectives were beginning to emerge on the causes of faunal change over time.

Despite various earlier rumblings, most modern accounts of how sci-ence finally came to grips with the changeability of life over the eons begin with the remarkable Jean-Baptiste de Lamarck. Working principally with mollusk fossils from rocks of the Paris Basin, this great French naturalist was the first to conclude explicitly—as early as 1801, and most influentially in his *Philosophie Zoologique* of 1809—that species, far from being fixed, modify as time passes. To Lamarck, species were genealogical lineages of organisms. Each lineage was discrete, had its own ancient origins, and pos-sessed innate tendencies toward change and greater complexity. And al-though this standpoint is quite distant from our ideas of evolution today, it did encapsulate the essential notion of change in the natural world, thereby representing a radical break with the traditional views of a static natural world derived from biblical scholarship. Perhaps it is unsurprising that La-marck was working in the secular environment of postrevolutionary France.

Given the profound implications of Lamarck's central insight for the history of life, it is tragic that it was—and continues to be—overshadowed by his unfortunate choice of mechanism: the notion that lineages modify via the use and disuse of their various anatomical features, as their members actively interact with their environments. A favorite example was the blind-ness of burrowing moles, although the most familiar is the giraffe's neck, stretched out from generation to generation as ancestral giraffes strove to feed ever higher in the trees. This notion of change is clearly wrong, for virtually none of the significant features an animal might acquire during its lifetime—a heavily muscled physique, for example, or flat feet—is directly or durably passed on to its offspring. But Lamarck's dynamic view does actually incorporate a further element—adaptation to the environment—that has been crucially important in later evolutionary thought. What is more, his view of change was an adaptation of prevailing beliefs based on the observation that environmental conditions might give rise to physical differences, such as the tanning of pale European skin in the tropical sun. Nonetheless, the unfortunate Lamarck has been singled out for ridicule ever since, with the result that the important babies of change and adaptation were rapidly thrown out with the bathwater of acquired-character inheritance.

In 1814, just five years after the appearance of Lamarck's great work, an Italian geologist named Giambattista Brocchi published a magisterial two-volume monograph on the rocks and marine fossils of the Apennine mountain chain in Tuscany. Like Lamarck, Brocchi tried to trace lineages of fossil organisms from successive strata. But although he, too, saw change among his fossils, he drew a very different conclusion. What he perceived was not a picture of steady transformation over time. Rather, the species that Brocchi identified in his fossil samples were relatively stable entities that—just like individual organisms—had births, lifespans, and extinctions. They appeared in the rocks; they persisted; they disappeared. And— again like individuals—they appeared to give rise to descendant offspring, in this case new species. By Brocchi's reckoning, lineages were not eternally separate, as Lamarck had thought: one could give birth to another! Brocchi soon moved on to other geological subjects, and most evolutionary biologists today would have difficulty recalling his name; but my colleague Niles Eldredge has argued persuasively that Brocchi's seminal ideas were an important influence on the young Charles Darwin, to whom we turn next.

WALLACE AND DARWIN

Charles Robert Darwin, probably the most influential biologist of all time, was born in 1809 (as it happened, the publication year of Lamarck's *Philosophie Zoologique*) into a world in which notions of mutability in Nature were already in the air not only in France, but also in his native Britain. Indeed, as far back as the late eighteenth century, Charles' own grandfather Erasmus Darwin had somewhat mystically mused on many of the features of the living world that his grandson later pondered. Here is a taste of those musings, buried deep within the first volume of Erasmus' lyrical medical treatise *Zoonomia*, published in 1794: "Would it be too bold to imagine, that in the great length of time, since the earth began to exist, perhaps millions of ages before the commencement of the history of mankind . . . that all warm-blooded animals have arisen from one living filament . . . endowed with animality, with the power of acquiring new parts, attended with new propensities . . . and thus possessing the faculty of continuing to improve by its own inherent activity, and of delivering down those improvements by generation to its posterity."

Prescient as these conjectures were, Darwin *grand-père* was far from the first observer to notice the patterns in nature that prompted them. According to the physicist and science historian Jim al-Khalili, almost a thousand years earlier the ninth-century Arab scholar Uthman al-Jahith, writing in Baghdad, had entertained ideas that were eerily similar to those Erasmus' grandson was to espouse: "Animals engage in a struggle for existence; for resources, to avoid being eaten and to breed. Environmental factors influence organisms to develop new characteristics to ensure survival, thus transforming into new species. Animals that survive to breed can pass on their successful characteristics to their offspring." This sounds almost too good to be true, but it does serve to show that the living world is clearly structured so as to provoke such speculations from the rare thinker unbound by received wisdom. Still, it remained true that in a conservative monarchy acutely aware of the recent violent events just across the English Channel, unconventional speculations of this kind had at the very least to be warily expressed.

Sent in 1825 to medical school at the distinguished but remote University of Edinburgh, the young Charles Darwin hated the blood and gore of surgery. But he obviously relished the intellectual energy of the "Scottish Enlightenment" that had made Edinburgh a hotbed of scientific as well as philosophical free-thinking. At the same time he began developing a serious interest in natural history, and fell particularly under the influence of two guardedly but avowedly Lamarckian scientists: the anatomist Robert Grant and the geologist Robert Jameson. Niles Eldredge believes that Jameson was the anonymous author of a pair of articles on the "transmutation" of species published in Edinburgh while Darwin was living there, and it is certainly rather improbable that Darwin didn't read these, or that he failed to see the short but appreciative obituary of Brocchi that appeared in the same journal in 1826.

As bracing as the intellectual environment of Edinburgh was, Darwin lacked the brash temperament required to stay the course in a blood-soaked, pre-anesthesia school of surgery. Accordingly, in 1828 (having learned taxidermy from John Edmonstone, a freed slave) he departed for the more socially and scientifically conventional environment of Christ's College, Cambridge. But it is pretty evident that he did not leave Edinburgh without thoroughly learning how to question received wisdom. Once at Cambridge, Darwin's path is well known. He became passionate about collecting beetles, which, even in temperate England, exist in an astonishing profusion of species. He regularly associated with the distinguished philosophers of science William Whewell and John Herschel. He went on an extended field trip to examine the rocks of Wales with the geologist Adam Sedgwick. Most important of all, he spent long hours in the herbarium with the botanist John Stevens Henslow. No avid young naturalist of the time could have hoped for a better grounding in the details of natural history. On the other hand, in the early nineteenth century faculty members at Cambridge were required to be ordained in the Church of England, and almost everyone with whom Darwin rubbed shoulders there would have been at least conventionally religious. So it was rather unlikely that there was much talk in Cambridge of the mutability of species—although Eldredge points out that Herschel, in his 1830 masterpiece *A Preliminary Discourse on the Study of Natural Philosophy*, did echo some of the views on the history of life with which Darwin would have been familiar from Edinburgh.

Still, after he graduated early in 1831, the 22-year-old Darwin was fully prepared for what was to be the most formative experience of his life. Through Henslow, he secured an invitation to accompany Robert Fitzroy, captain of the British Navy sloop *Beagle*, on a mission to map the coastline of South America—an enterprise that eventually turned into a five-year, around-the-world epic. The high points of this voyage have been endlessly recounted: how discoveries along the way made Darwin aware of the astonishing diversity of different species across the world, especially in the tropics; how strongly impressed he was by finding fossils, near the Brazilian port of Bahia Blanca, of extinct armored glyptodonts that were supplanted in time by their living relatives the armadillos; how he noticed that different kinds of rhea (ostrich-like flightless birds) similarly replaced one another on the landscape as he journeyed south through Argentina; how horrified he was by the barbarities of the Brazilian colonial and slave systems; and most famously, how he saw that the mockingbirds of the Galapagos archipelago, while related to those on the South American mainland, differed themselves from island to island.

In his book *Eternal Ephemera*, Eldredge makes a strong case for concluding that by the end of his first year at sea Darwin was already mentally weighing the cases for Lamarckian transformation versus Brocchian replacement in producing the profusion of living species that so overwhelmed him everywhere he looked. And Eldredge points out that, even if Darwin had improbably contrived not to learn about Brocchi's ideas in Edinburgh, he certainly would have known about them once he had received the second volume of the geologist Charles Lyell's great *Principles of Geology* while the *Beagle* was docked at Montevideo in November 1832. For even as Lyell expressed disapproval of Brocchi's views of transmutation—and of Lamarck's transformationism, for that matter—he expounded both with great clarity in a volume that Darwin eagerly devoured.

Whatever the exact case it is evident that, by the end of the *Beagle's* voyage in late 1836, Darwin had become convinced that species were mutable so that all species in the living world must have arisen from relatives now extinct. This much was clear to him. But was the transformation seen among the fossils gradual, as Lamarck had believed, or a matter of replacement, as Brocchi had suggested? And in either event, how did the transformation come about? These were tougher questions over which Darwin privately agonized for years, although there were early signs that he would come down firmly on the side of gradual change.

Meanwhile there was a flurry of activity, as the newly famous young naturalist sorted out the collections he had made on his voyage, distributed them among various specialists for study, and began to organize his thoughts. By the middle of 1837, Darwin had drawn the first hypothetical evolutionary tree in a private notebook, confirming his belief that old species give rise to new ones. In 1842, having read a late edition of the English cleric Thomas Malthus' celebrated tract on human population growth ("which, unchecked, goes on doubling itself every twenty-five years"), he produced a quick "Sketch" of his developing ideas on evolution by natural selection. Two years later he fleshed these out into a lengthy *Essay* that was never published in his lifetime, though he did leave instructions for its publication in the event of his death.

Mature Darwinian reasoning goes as follows. The pattern of resemblances seen in the living world (those "nested sets" that Linnaeus had noticed) is best explained by "descent with modification" as, over vast eons of time, living things diverged repeatedly from a single ancestral form. Highly dissimilar organisms, such as birds and sponges, are descended from a very ancient common ancestor, while very similar ones, such as different species of antelope, shared an ancestor quite recently. This divergence comes about because individuals in any population vary in their inherited features, and many more individual organisms are born into a species than ever survive to reproduce. The successful reproducers are the individuals whose heritable features best adapt them to flourish in the environment—which thus exerts a consistent pressure toward physical change from generation to generation, as the better-adapted consistently out-reproduce their less favored brethren. In this way, Nature fine-tunes each population to its habitat; as its adaptation improves, its aspect changes. Such "natural selection" is analogous to the artificial selection used so efficiently by animal breeders since time immemorial to obtain stock with desired characteristics, the difference simply being that in the case of natural selection there is no goal in view: the force involved is immediate exigency. As a corollary of this generation-by-generation process, species evolve gradually, transforming themselves from ancestor into descendant—essentially as envisaged by Lamarck—as differences accumulate under the unconscious guiding hand of natural selection.

The basics of this argument were already there in Darwin's *Sketch* of 1842, in which he had come down on the side of transformation, rather than replacement; it was pretty fully stated in the 1844 *Essay*. Put all the elements together, and they make a coherent and magnificently reductionist story that many still find highly compelling. But even when backed by a specific mechanism of change, the softer-edged transformationist notion nonetheless cut cleanly across mid-nineteenth-century assumptions about Nature. And while he was acutely aware of the radical importance of transformation, Darwin was also very conscious of the fact that his evidence for it was not conclusive, at least by his own high standards.

As a result, he was hugely reluctant to go public with his theory, for any one of many possible reasons. These certainly included his reluctance to upset his devoutly Christian wife, who had not wished to marry anyone she might not be reunited with in Heaven; his own chronic ill health; and his inability to bear the thought of the scientific uproar he knew would follow publication—though maybe the proximate cause of his hesitation was something else entirely. But whatever the underlying reason, Darwin's scientific trajectory in the decade following 1844 looks very much like what a modern ethologist would call "displacement" activity (of the kind a subordinate male macaque shows, for example, when he chases a juvenile instead of challenging the alpha male for the receptive female he truly desires). For almost a decade following his completion of the 1844 manuscript, Darwin threw himself into an immensely detailed study of sessile marine invertebrates, something he did with such manic dedication that one of his young sons, visiting a friend, is said to have asked, "Where does *your* father do his barnacles?" The four resulting volumes earned Darwin a solid scientific reputation and the Royal Society's most prestigious medal, while meticulously documenting variation in a whole slew of species; nonetheless, they showed no trace of the revolutionary thinking that he eventually confided privately to a very small circle of trusted naturalist colleagues, in one case with muffled guilt, "as if confessing a murder."

While the affluent Darwin was mulling his evolutionary ideas in the comfort of his country home, his younger contemporary Alfred Russel Wallace was enduring incredible hardships on the other side of the world. Every bit as keen a naturalist as Darwin, and anxious to see as much of nature as possible, the perennially impecunious Wallace left England in 1848, at the age of 25, to collect natural history specimens in Amazonia. He did this for his own edification as well as to make a living from the sale of the specimens he collected; he made his foray to the tropics with the "species question" very much in mind. During the return voyage to England from Brazil in 1852 his ship caught fire, resulting in the loss of virtually all of his collections and a perilous fortnight adrift in the mid-Atlantic in a leaky lifeboat. Undeterred, he left England again in 1854, this time bound for the Malay Archipelago, the vast chain of islands lying between Singapore and New Guinea. Living in the most primitive of conditions, and constantly dependent on the kindness of strangers—who were often native peoples who'd had little previous contact with outsiders—he industriously collected thousands of specimens and made extensive notes on everything he saw, drinking in the tropical luxuriance of animals, plants, and cultures. And from the beginning, as he had in South America, he asked himself how this profusion of species could possibly have come about.

In early 1855, while collecting in Sarawak, northern Borneo, Wallace clearly revealed the direction of his thinking by sending back to England a manuscript cogently titled *On the Law Which has Regulated the Introduction of New Species.* In this he roundly declared that "every species has come into existence closely coincident in space and time with a closely allied species," a statement that he seems to have formulated specifically in opposition to Lyell's rejection of the ideas of Lamarck and Brocchi. Lyell read this paper when it was published several months later and was evidently powerfully affected by its reasoning and evidence. He showed it to Darwin on an 1856 visit, coincidentally the very occasion on which he became only the second colleague to whom the latter confided his evolutionary thoughts. Darwin had by that time already begun a correspondence with Wallace, to whom he wrote to assuage Wallace's fears that his paper had been ignored—a generous gesture that may account for what happened next.

The beginning of 1858 found Wallace collecting on the island of Ternate, in the remote South Molucca islands. He had already assembled a large body of evidence in favor of transmutation, and at that point all he lacked was a mechanism whereby it might happen. Suddenly, he had it. During one of the many bouts of fever that wracked him during his sojourn in the East, he somehow put it all together: evolution by natural selection. When his strength returned, he immediately wrote a manuscript provocatively titled *On the Tendency of Varieties to Depart Indefinitely from the Original Type.* Knowing nothing of what was transpiring back in England, he sent it to his kindly correspondent Darwin, with the request that he show it to Lyell if he considered it "sufficiently important."

Important? Darwin was devastated. Shortly after receiving the manuscript he wrote to Lyell: "I never saw a more striking coincidence. If Wallace had my . . . sketch written out in 1842 he could not have made a better short abstract! Even his terms now stand as Heads of my Chapters." And then the *cri de coeur:* "All my originality, whatever it may amount to, will be smashed." In the event, Darwin's fears were not realized. Lyell and other colleagues arranged for some of his writings to be presented jointly with Wallace's paper at the Linnean Society of London; and once this had been done Darwin set feverishly to work, leaning heavily on his 1844 *Essay* to speedily produce his fat and fateful volume *On the Origin of Species by Means of Natural Selection.* This masterwork was completed in time to become a publishing sensation at the end of 1859, and, as Darwin had feared, the outcry was instant and scandalized (one society matron's exclamation: "Descended from the apes? My dear, let us hope that it is not true; but if it is, let us pray that it does not become generally known," has become a classic). Nonetheless, in many ways it is remarkable how quickly straitlaced Victorian society became used to the idea that the evident unity of life is due to common descent.

Knowing how much uproar his ideas would cause, the introverted Darwin deliberately avoided making matters worse by broaching their implications for humankind. In the *Origin* he contented himself with the cryptic phrase "Light will be shed on the origin of man and his history," preceded by the prediction, ignored at the time but seized upon in our day by evolutionary psychologists, that "in the distant future . . . psychology will be based on a new foundation, that of the necessary acquirement of each mental power and capacity by gradation." But despite his discretion, public reaction was forceful enough to send Darwin right back into displacement mode, even as he clearly recognized that in some way he eventually would have to take the bull by the horns. Accordingly, over the decade following the *Origin's* appearance, the industrious Darwin turned his energies to producing substantial volumes on orchids and the domestication of plants and animals. Thus it was only in 1871, when the fuss over evolution had subsided, that he came out with his opus *The Descent of Man, and Selection in Relation to Sex.*

Even then, this hefty two-volume work was not quite what its title promised. Most of its first volume consists of Darwin's refutation of the polygenist theory of the human races: the notion, exploited by the supporters of slavery, that the geographic divisions of humankind had been separately created—or latterly, following Darwin's own evolutionary notion, that they had evolved from different ape species. Darwin came from a prominent family of abolitionists, and during the voyage of the *Beagle* he had been as profoundly horrified by the mistreatment of Brazilian slaves as he was enchanted by the wonders of Nature. And he almost certainly wrote the *Descent* less as a logical extension of the thoughts expressed in the *Origin* than out of a sense of obligation to refute the claims of the polygenists and other apologists for slavery. Still, the work seems to have taken on a life of its own as its writing progressed, and the bulk of it was eventually devoted to an exposition of Darwin's notion of sexual selection. This is the idea that mate choice is an important influence in evolution, the classic example being peacocks, which appear to have cumbersome giant tails simply because peahens prefer them that way. Perhaps this focus on sexual selection was inevitable since, just as with evolution itself, the case for monogeny—the single common origin for all humans that Darwin stressed—was inextricably bound in his mind to its underlying mechanism. And because Darwin was not much impressed by any adaptive differences among the races, it was sexual selection—basically, differing standards of beauty—that became his chosen mechanism to explain "the divergence of each race from other races, and all from a common stock."

To the extent that the *Descent* delivered on its main title at all, it did so only indirectly, by exploring continuities in anatomy, and especially in behavior, between humans and other animals. What it conspicuously did *not* consider was the issue of human descent itself. This was at the very least odd because, by the time the *Descent* appeared, humankind did indeed have a historical record, if a highly limited one. Most significant among the tiny handful of human fossils known was a partial skeleton from the Feldhofer Grotto in Germany's Neander Thal (Neander Valley) that had actually been discovered in 1856, before even the *Origin* had been published. This fossil had made a huge splash in British scientific circles when its description was translated into English in 1861; and yet a decade later, in the entire two volumes of the *Descent*, the distinctive and clearly very old Neanderthal skull cap merited just one single mention: a throwaway line in which Darwin observed that even ancient humans could have large brains.

This is all the more amazing because, as early as 1864, Darwin had seen—and presumably handled—a second Neanderthal skull. Discovered at some time before 1848 in the British territory of Gibraltar—where it had lain on a shelf in a local museum until it was spotted by a visitor and brought to London—this remarkable specimen was described by the

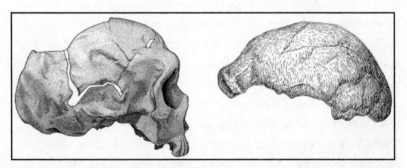

Two of the first hominid fossils ever discovered. Left: the Gibraltar cranium examined by Darwin in September 1864. Right: the Feldhofer Neanderthal fossil, type of Homo neanderthalensis. *To scale. Drawn by Don McGranaghan (left) and Diana Salles.*

English anatomist George Busk in mid-1864. On September 1 of that same year Darwin wrote, in a letter to his close colleague J. D. Hooker, that he had been visited at his London residence by Lyell and the paleontologist Hugh Falconer. At the end of this missive, he offhandedly mentioned that "F brought me the wonderful Gibralter [sic] skull." And that, as far as I have been able to discover, was the only mention Darwin ever made of this specimen—or, effectively, of any hominid fossil. By the time he published the *Origin* Darwin had evidently developed severe reservations about the incompleteness of the fossil record, which inconveniently failed to reveal the slow, insensible transitions his theory predicted. As time passed these reservations increased, so whereas he had been prepared to speculate about fossil apes and possible human ancestry in his youthful post-*Beagle* notebooks, by midcareer his attitude toward the hominid fossil record had become downright evasive. Wonderful as the Gibraltar hominid might have been, the human fossil record was a place Darwin just didn't want to go.

While Darwin was busily becoming the most famous biologist in history, Wallace—of whom someone once said that he was most famous for not being famous—was continuing his labors in the Malay Archipelago. He remained there until 1862, eventually amassing a collection of more than 126,000 specimens (most of them beetles) that contained many species new to science. When he returned to England he found his professional circumstances greatly improved by his new association with Darwin and other eminent scientists—though he never found a paying job appropriate to someone of his reputation, and lived in precarious financial circumstances for the remainder of his life. And although Darwin had been thoroughly shocked by the similarity of their evolutionary ideas when he initially read the Ternate manuscript, the views of the two naturalists actually differed in ways that were bound eventually to become obvious.

The divergence in perspective most often remarked upon is that by the time he wrote the *Origin* Darwin had come to see the individual struggle for reproductive success as the driver of change, while Wallace was always at least as concerned with the fates of what he called "varieties." At the risk of oversimplifying, one could say that for Darwin the basic unit of selection was the individual organism, while for Wallace it was as much the local population, or even the species. Accordingly, while Darwin could logically see species as ephemera that evolved themselves out of existence, in Wallace's world species had a more solidly Brocchian presence.

There were also other significant matters on which the two naturalists saw things differently. For example, Wallace was never much enamored of Darwin's idea of sexual selection. But their most famous disagreement came specifically on the issue of human evolution. However reluctant he may have been to confront actual human fossils, Darwin was clearly content to explain all the peculiarities of modern human beings as the products of natural selection. No special pleading required. Wallace, in contrast, had extreme difficulty in perceiving how natural selection could have driven into existence humanity's most vaunted features: its large brain and the unique form of consciousness that flowed from it. Wallace saw that selection stemmed from immediate needs, but he also felt strongly that the modern human brain was a lot bigger than the most primitive human lifestyles (including the ancestral one) demanded. Yet all humans had big brains. How could selection have provided something so much in excess of requirements?

Wallace was in no way conventionally religious, but he had his spiritual side, and this reasoning process drove him to ask whether some form of "Overruling Intelligence" might have given the laws of Nature a nudge in this case. Darwin was predictably aghast and wrote to Wallace, "I hope you have not murdered too completely your own and my child." This rift between the coauthors of evolution by natural selection eventually healed, although perhaps never completely: there was one famous instance in which Darwin and Wallace found themselves testifying on opposite sides in the judicial prosecution of a spiritualist for fraud. And despite all that we have learned since, the Darwin-Wallace dispute over human evolution reflects a dichotomy of views on the origin of human consciousness—gradual or abrupt—that, as we'll see, still resonates in modern paleoanthropology.

CHAPTER 2

PEOPLE GET A
FOSSIL RECORD

ONE POSSIBLE—IF ON ITS OWN HARDLY CONVINCING—REASON why Darwin did not directly broach the issue of fossils and human ancestry in *The Descent of Man* was that his friend and vigorous defender Thomas Henry Huxley had already treated the subject at some length in early 1863, a scant three years after *On the Origin of Species* had appeared. Huxley was an accomplished anatomist who later won paleontological fame when, allegedly while carving his family's Christmas fowl, he realized that birds are related to theropod dinosaurs—a notion that is finally back in fashion. In his influential 1863 series of essays, *Evidence As to Man's Place in Nature*, Huxley predictably enough brought a comparative anatomist's eye to the issue crisply summarized in his title. Given this focus, he was principally concerned, as Darwin was to be in the *Descent*, with substantiating Linnaeus' classification of humans as yet another mammal species, something he convincingly accomplished by minutely documenting continuities of structure between humans and other animals, most particularly the African apes. But he devoted one very influential essay specifically to "Some Fossil Remains of Man."

The fossil remains in question were effectively three. Two of them, the braincases of an adult and a child (a third cranium has disappeared), had

been discovered at a Belgian cave called Engis by the physician and antiquarian Charles-Philippe Schmerling in around 1830. Schmerling was an
astute observer, and far ahead of his time in recognizing that the human remains were associated with the bones of the extinct mammoth and woolly
rhino, and thus truly ancient; but before Huxley picked up on them, few
had paid them much attention. Huxley's third hominid fossil, which effectively completed the scientifically useful hominid record known at the time
(the Gibraltar skull had yet to be announced), was the adult Neanderthal
skullcap from the Feldhofer Grotto.

The Engis fossils appeared clearly ancient, having been found apparently sealed into ancient sediments alongside those extinct Ice Age mammals; and in Huxley's opinion, the Feldhofer Neanderthal was also of "great,
though uncertain, antiquity." He dilated at length about the Engis adult ("a
person of limited intellectual faculties"), an unfortunate focus because we
now know that the skull belonged to a modern *Homo sapiens* who had been
buried into the cave's Ice Age deposits a mere 8,000 years ago. Ironically,
the immature cranium, about which Huxley said little—probably because
it lacked much of the characteristic bony anatomy present in the adult—has
subsequently been identified as a juvenile of the species represented at the
Feldhofer Grotto. But Huxley's choice left him with the Feldhofer fossil as
the only truly archaic specimen to discuss.

That historic fossil had been accidentally discovered in 1856, two years
before Darwin and Wallace went public with their evolutionary ideas.
Seeking lime to fuel the burgeoning chemical industries of the Rhineland,
quarrymen had been digging down into rubble deposits within the Feldhofer cave. The rubble, containing the remains of extinct cave bears, had
to be cleared out to expose the desired pure lime, and the miners had energetically shoveled most of what had probably been a complete hominid
skeleton out of the cave entrance, and down the cliff below, before what
was left was saved by an alert supervisor. These few bones came into the
possession of a local schoolteacher named Johann Fuhlrott, who to his
eternal credit immediately and correctly divined that they were human,
unusual, and ancient—although he later contrived to write an entire book
about them without ever broaching the issue of what kind of human they
represented! Finding himself out of his depth with this unprecedented
skeleton, Fuhlrott quickly passed it along for analysis to the distinguished
Bonn anatomist Hermann Schaaffhausen.

Equally lacking any appropriate intellectual framework within which to interpret the fossils, Schaaffhausen seems to have been just as bemused as Fuhlrott was. He recognized that the heavily fossilized Feldhofer skeleton was old, but bound by conventional beliefs he also doubted its contemporaneity with the "antediluvian" animals found at the site. So his conclusion was almost inevitable that, despite its peculiarities, the Neanderthal fossil had belonged to a member of an ancient "barbarous race" of *Homo sapiens* that had inhabited Europe in ancient times. Still, after comparing the skullcap to a whole variety of human crania that were also considered ancient (whatever that might have meant; at the time he certainly couldn't even have dreamt that, as we know today, the Neanderthaler is 40,000 years old), Schaaffhausen nonetheless found it necessary to emphasize that the human bones showed "a natural conformation hitherto not known to exist," and that they "exceed all the rest in those peculiarities of conformation which lead to the conclusion of their belonging to a barbarous and savage race." In other words, he clearly recognized that the Neanderthal was *different*.

And indeed it was. Just for a start, although the Neanderthal skullcap had contained a brain as large as that of a modern person, it contrasted strikingly with our high, rounded, and delicate braincases in being thick-boned, long and low, and in retreating toward a protrusive rear. What's more, at the front it was decorated by a pair of heavy, prominent browridges that arced gracefully over each eye socket. You may occasionally see a modern human sporting fairly pronounced eyebrow ridges, but never anything like this.

Of all people scientifically active at the time, you might have expected Thomas Huxley, who described these features at some length, to have been the first to appreciate their significance. This is not only because Huxley was an enthusiastic evolutionist who had berated himself with the words "Why didn't I think of that?" after hearing the historic presentation at the Linnean Society in 1858, and who had gone on to become Darwin's most vociferous public defender. It is also because, unlike the gradualist Darwin, who might legitimately have been thrown off the scent by the large brain of the Feldhofer individual, he was also a Brocchian "saltationist."

On June 25, 1859, Huxley penned a letter to Lyell—who was much more reluctant than he was to accept evolution by natural selection—that was very clear about how he, Huxley, thought evolution worked. "The fixity and definite limitation of species, genera, and larger groups appear to me to be perfectly consistent with the theory of transmutation," Huxley wrote. "In other words, I think *transmutation* may take place without transition . . . and in passing from species to species 'Natura facit saltum.'" Huxley's Latin aphorism (meaning "nature makes jumps") was an ironic reversal of the ancient mathematical axiom *natura non facit saltum:* a guiding principle of natural philosophers since Aristotle's time that had been enthusiastically embraced by Darwin—misguidedly, in Huxley's view. "My dear Darwin," he wrote in an otherwise highly laudatory letter, written on November 23, 1859, just after the *Origin* had appeared, "you have loaded yourself with an unnecessary difficulty in adopting *natura non facit saltum* so unreservedly."

Huxley's own theoretical stance was thus clear: species are discontinuous. So if, in good Brocchian fashion, Nature made jumps between species, and if the Neanderthal was as idiosyncratic as Schaaffhausen had claimed it was—and as his own descriptions had fully substantiated—you might have expected Huxley to reach the obvious conclusion. Namely, that the Feldhofer Grotto had yielded evidence of a new human species, now apparently vanished from the world: a species that was clearly related in some way to *Homo sapiens,* but that was at the same time distinct. In a newly evolutionary scientific world, in which geologists had already known for many years that species are routinely supplanted in time by their close relatives, this would hardly have been an unacceptable deduction. And while it would doubtless have been derided in the popular press, that prospect would hardly have deterred the combative Huxley, already widely known as "Darwin's bulldog." But for reasons we can still only imagine, Huxley chose not to go this way. Instead, through some breathtaking intellectual legerdemain, he eventually arrived at the conclusion that "though the most pithecoid of human skulls, the Neanderthal cranium is by no means so isolated as it seems at first, but forms, in reality, the extreme term of a series leading gradually from it to the highest and best developed of human crania."

One way in which Huxley demonstrated this alleged continuity was by comparing the contours of the Neanderthal braincase to those of an Australian aborigine skull, which, in fine Victorian fashion, he designated the most primitive form of *Homo sapiens.* In doing this, he was very clearly harking back to a long-established tradition that hierarchically organized the various geographical varieties of humankind into an ascending series with Europeans at the pinnacle. And while Huxley's scientific conclusions

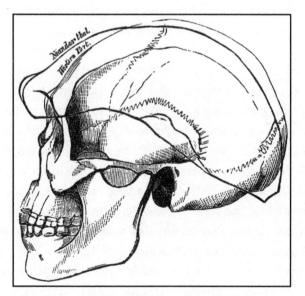

Huxley's contrast of the Feldhofer Neanderthal skullcap's outline with that of a modern Australian skull, in his Man's Place in Nature *of 1863. See text.*

comfortingly reflected the social prejudices of his time, it is hard to imagine anything he could have done that would have been more antithetical to the evolutionary view that he had otherwise so enthusiastically embraced. For this implicit ranking of human races represented nothing less than a return to the medieval Great Chain of Being, the very mind-set from which Darwinian thought offered the best hope of escape.

So here we have, at the very beginning, a (literally) textbook example of the exceptionalism that has benighted the science of paleoanthropology ever since. Had Huxley not been dealing with a creature that was so obviously in some way human he would, without any doubt whatsoever, have acknowledged that the Neanderthal fossil represented an independent and previously unknown entity. After all, his fellow paleontologists were busily describing new extinct mammal species all the time. But, while he did fleetingly consider the possibility that the Neanderthal bones could "be regarded as the remains of a human being intermediate between Men and Apes," Huxley summarily rejected this notion in favor of the view that "a small additional amount of flattening and lengthening, with a corresponding increase in the supraciliary ridges, would convert the Australian brain case into a form identical with that of the aberrant fossil."

A quick look at Huxley's illustration (see figure, page 31) is enough to show that you would have to do a whole lot of pushing and squeezing and distorting to make the Australian into a Neanderthal. Still, this small difficulty was evidently not going to prevent Huxley from performing the necessary mental gymnastics. He pulled off the trick by displaying the traditionalist human anatomist's hallmark: an obsessive concern with variation: with the differences that exist among individuals of our single species. But while individuals do indeed differ among themselves, in anatomy just as in temperament and behavior, the differences you typically see *within* a mammal species, *Homo sapiens* included, tend to be variations on a single theme. And the anatomical peculiarities of the Feldhofer fossil, compared to, say, the modern Engis adult, are clearly differences of theme itself. They speak of identity, and of diversity: of the differences you typically see *among* related species.

There are plenty of instances in the extensive human fossil record we now have at our disposal in which it is really tough, or even impossible, to judge where intraspecies variation ends and interspecies variation begins. But the case of the Neanderthal is emphatically not among them. The Feldhofer hominid and its kin really *are* anatomically distinctive when compared to modern humans, as Schaaffhausen recognized. But while Huxley was willing to admit in principle that in ancient times human forms must have existed that were biologically intermediate between "primitive" ancestors and "advanced" descendants, he refused—exactly like his colleague Darwin—to identify any tangible fossil as such. And in this refusal, which entailed arbitrarily cramming something truly distinctive into an already-existing species, he was unwittingly founding a tradition that has bedeviled paleoanthropology ever since.

The first formal suggestion that the Feldhofer remains represented a hominid distinct from *Homo sapiens* was made shortly after *Man's Place in Nature* was published. But meanwhile, as the very first human fossils to come to scientific attention, the German hominid had fittingly and immediately unleashed the very first major controversy in what has become the notoriously quarrelsome science of paleoanthropology. Barely had Schaaffhausen pronounced his conclusions about the curious new find at a meeting of the Lower Rhine Medical and Natural History Society in 1857 than he found himself vigorously assailed not only by his fellow Bonn faculty member August Franz Mayer, but also by the pathologist Rudolf Virchow, the most famous medical man of his day. These worthies both contended that the peculiar anatomy of the Neanderthal individual did not reflect the characteristics of the population to which it had belonged, as Schaaffhausen had insisted it did. Instead, it was due to disease that had altered the skeleton during the individual's lifetime (pretty much as the amazing tiny skeleton found in the Indonesian island of Flores in 2003 was, almost a century and a half later, also said by many to be pathological).

Because of his high scientific reputation, Virchow's view—born of an ironclad conviction that species were immutable—that the Neanderthal remains were those of an unfortunate modern human who had suffered the deficiency disease of rickets in childhood, cranial injury in middle age, and arthritis in senescence, carried greater weight. But Mayer's more specific interpretation—to which Virchow gave his imprimatur—was significantly more extreme. The physician's appraisal of the skeleton indicated to him that the preserved leg bones were curved by both rickets and a life on horseback. The pain of his affliction had caused the unfortunate individual to chronically furrow his brow in agony, leading to the excessive development of bone above the eye sockets. Here were the elements of high drama! Denying the antiquity of the bones, Mayer concluded that the Feldhofer remains were those of a Cossack soldier belonging to one of the Russian armies that had rampaged across Germany in 1814 on their way to attack France. After being wounded, this unfortunate individual had crept into the Feldhofer cave to die—and then had implicitly contrived to become covered, in a mere half century, by two meters of fossil-bearing sediments.

Huxley, predictably enough, had a lot of fun with this absurd scenario. But he nonetheless could not bring himself to think of the Neanderthal as anything but an extreme example of modern human variation. Not so William King, an Anglo-Irish geologist who, remarkably enough, had studied with the reluctant evolutionist Charles Lyell. At a scientific meeting in 1863, King proclaimed that the Feldhofer fossils represented a distinctive human species, *Homo neanderthalensis*; he followed up this verbal declaration with a formal publication the next year. He wrote of the "remarkable absence" in the Feldhofer skullcap of "those contours and proportions which prevail in the forehead of our species." So notable was this absence that "few can refuse to admit that the deficiency more closely approximates the Neanderthal fossil to the anthropoid apes than to *Homo sapiens*." Having gone this far, it was only a short step to King's dramatic conclusion that, large as the braincase might have been, the "thoughts and desires that once dwelt within it never soared beyond those of the brute."

Using somewhat more dispassionate terminology, paleoanthropologists still debate King's last claim today. But King's perception that the Neanderthal fossil was indeed distinctive was substantiated, almost immediately, by George Busk's announcement of the Gibraltar cranium that Darwin was shortly to see. Here was a more complete specimen that preserved an impressively large face and that, though somewhat more lightly built, resembled the Neanderthal specimen in comparable respects. The similarity was so striking that "even Professor Mayer," Busk wrote, "will hardly suppose that a rickety Cossack engaged in the campaign of 1814 had crept into a sealed fissure in the Rock of Gibraltar." Clearly, the Feldhofer individual was no pathological flash in the pan.

Or so one might have thought everyone would conclude. But in reality the presentation to the world of the Gibraltar Neanderthal did little to settle the argument, at least in the short term. This was particularly true in Germany, where the immensely influential Rudolf Virchow was still arguing in 1872, the year after *The Descent of Man* appeared, that the Feldhofer individual had been so severely afflicted by his various maladies that he would have been unable to survive in an ancient pre-agricultural society, and thus had to be recent. And in England even the redoubtable Busk had concluded, in good Huxleyan fashion, that although his Gibraltar skull was "very low and savage," it was "still man, and not a half way step between man and monkey." Substantiation of *Homo neanderthalensis* as a distinctive biological phenomenon—proof that it was something that needed to be confronted on its own terms, rather than explained away—had to await discoveries that did not come for another 20 years.

THE ANCIENTS

Meanwhile, the antiquarians had been hard at work. The flint tools that were regularly turned up by ploughs in fields all over Europe had long been recognized as ancient, and in England the antiquarian John Frere had remarked as early as 1800 that chipped flint tools unearthed deep in a quarry may have belonged to "a very remote period indeed, even beyond that of the present world." In other words, they were deliberately made implements that dated from before Noah's Flood. Thirty years later, across the Channel in northern France, a customs official called Jacques Boucher de Perthes became similarly convinced that stone tools found in abundance in the terraces of the Somme River were truly ancient. For many years his opinion fell on deaf ears, until a visiting delegation of distinguished British geologists validated it in 1859, the year of the *Origin of Species*. Their visit to the sites at issue had been inspired by the same Hugh Falconer who later brought the Gibraltar fossil to Darwin.

From midcentury on, great strides were made on the European continent in establishing that our species *Homo sapiens* has been around for a long time. As early as 1852 excavation of a rock shelter at Aurignac, in the foothills of the French Pyrenees, had produced *H. sapiens* skeletons in association with obviously ancient extinct animals and stone tools; and by the 1870s, fledgling archaeologists were beginning to develop a rough relative chronology for the succession of cultures that had existed in Europe over the late part of the last Ice Age. That chronology was based almost entirely on the excavation of archaeological deposits that had accumulated in rock shelter and cave entrance sites in which ancient humans had lived, and it was relative because at the time it was only possible to determine what was older or younger than something else: year dates were still a century in the future.

Just as in the geological record, deposits build up over the years at human occupation sites, with the oldest layers at the bottom of the heap. Sterile layers consisting of dust and rock fragments, accumulated during times when the sites were uninhabited, are interspersed with layers containing artifacts and animal bones left behind by ancient human occupants. Examination of the contents of the occupation layers showed that the exact nature of the stone tools and the species of animals typically changed from one such layer to the next. Those changes were the key to developing the cultural chronology of each site and region.

By 1872, the French archaeologist Gabriel de Mortillet had begun roughing out a sequence of cultures over the Paleolithic, the "Old Stone Age." Each culture was named for a "type site" and was recognized from the kind of stone implement that characterized it. As information accumulated the sequence became more complex, and more periods were recognized. Oldest in Mortillet's sequence as eventually developed was the Acheulean, named for St. Acheul, one of Perthes' sites. It is typified by the large tear-drop-shaped implements known as hand axes, produced by banging multiple small flakes off suitable rock cores. Above this lay the Mousterian, characterized by small triangular hand axes, scrapers, and points usually made on flakes. Next up in the pile came the Aurignacian. This differed strikingly from the Mousterian in frequently employing antler and bone in addition to rock as raw materials for implements; indeed, the diagnostic Aurignacian tool is a slender bone point with a split base. Flint tools in the Aurignacian are typically long, thin "blades" that were struck in sequence from a cylindrical stone core. After the Aurignacian came the Gravettian culture, typified by the straight-sided burin with a narrow triangular point, and this was followed by the Solutrean, most immediately recognizable from astonishingly delicately worked flint "laurel leaves" fashioned on long blades. Almost at the top of the Old Stone Age sequence was the Magdalenian, with a wonderful diversity of implements, including tiny microliths that were designed to be mounted in handles. The Magdalenian faded away at the end of the Ice Age as the polar ice cap retreated for the last time, to be replaced by materially simpler Epipaleolithic and Mesolithic cultures that, after a long lapse, yielded to the Neolithic cultures of the earliest farmers in the region.

Most of the habitation sites investigated by the early archaeologists contained evidence of more than one Paleolithic culture. But they always occurred in the same order, providing the basis for a chronology that survives basically intact to this day, although it is now calibrated with year dates. The Acheulean is very early, fading out by about 250,000 years ago, which is about when we find the first signs of the Mousterian, the principal culture of the Middle Paleolithic. The latter persisted until about 30,000 years ago in western Europe, having been joined by the Aurignacian, the first culture of the Upper Paleolithic, at about 40,000 years ago. The Gravettian took over around 28,000 years ago, yielding at around 22,000 years ago to the Solutrean, which in turn gave way to the Magdalenian at about 18,000 years ago. The last Ice Age itself began to end about 12,000 years ago, the Magdalenian yielding to the Mesolithic shortly thereafter.

Each of these "cultures" is defined by a particular way of working stone, but archaeologists understandably take these narrow differences as proxies for a much wider range of activities that embraces most of the elements we would now use to describe culture in its broader sense. Today it is pretty clear that the pattern over time is one of cultural replacement, rather than the steady evolution of one culture into another; but back in the late nineteenth century that was not the view of the politically progressivist Mortillet. Not only did he perceive an intergradation of cultures, but in good Darwinian fashion he was rather explicit about expecting to find a graded series between an apelike ancestor and modern humans. Still, like Darwin's, these intermediates remained purely hypothetical; and together with Huxley and Busk, Mortillet viewed the Neanderthal fossil not as a separate kind of being but as a primitive human. It was not quite at the pinnacle like a modern European, but it was close enough to be in the same species. Transformation there was, but transmutation there wasn't.

Thus, throughout the 1860s and 1870s, the species *Homo neanderthalensis* continued its ghostly existence in an uneasy no-man's-land of paleoanthropology, even as bone fragments that were sooner or later to be attributed to it continued to turn up at sites scattered across Europe. But nothing was found to make much fuss about until 1866, when two spectacular skeletons were discovered in undisturbed archaeological deposits at a cave entrance near Spy, in Belgium. Both skeletons were reasonably complete; both were

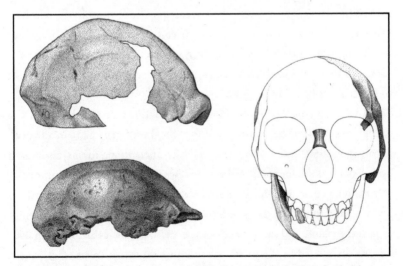

Three of the hominid fossils mentioned in this chapter. Top left: Neanderthal from Spy, Belgium. Bottom left: the Trinil skullcap from Java, type specimen of Homo erectus. *Right: one reconstruction of the fraudulent "fossil" from Piltdown, England. To scale. Drawn by Don McGranaghan.*

ancient; and both were clearly associated with Mousterian tools. What's more, feature for feature they matched the hominids from Feldhofer and Gibraltar. At last, here was evidence of a distinctive Neanderthal identity that was compelling enough to make almost everybody pay attention. No pathological oddities these. For the first time, fossils were available that, for anyone prepared to look carefully, conveyed a very good idea of what an entire Neanderthal was like: a Mousterian tool-wielding, heavy-boned, and large-brained human relative of modest stature, with a big, jutting face and a low cranial vault that retreated behind bulky browridges.

JAVA

The year after the Spy discovery, a young Dutch anatomist named Eugene Dubois set sail for Wallace's Malay Archipelago with the avowed intent of finding the fossil remains of early man. There is not much in his rather conventional medical school background to explain why he should have embarked on this quixotic quest, but there is no doubt that Dubois was an avid Darwinian, and that he was determined to bestow a material identity on the hypothetical human precursors about which Darwin and his followers had written at length. Convinced that the Neanderthals were pathological, and that modern human beings had evolved outside Europe, he began looking toward the tropical homes of today's great apes, our closest living relatives. Darwin's preference had been for an African ancestry of humankind, but for the impecunious Dubois it was far more practical to reach the tropics in what was then the Dutch East Indies, as a medical officer in the employ of the Dutch army.

The region had several additional attractions. It was home to the orangutan, then the best known of the great apes; fossils had already been found there that were said to resemble the ancient ape fossils discovered by Falconer in the foothills of the Himalayas; and the military authorities seemed ready to assist Dubois in his eccentric endeavors. He started out by investigating limestone caves in Sumatra, but in 1890 he shifted his attention to the island of Java. Almost immediately, at a site called Kedung Brubus, he discovered his first truly ancient human fossil: an enigmatic piece of jaw containing one tooth. But it was in the next year that he hit real pay dirt not far away, at a place called Trinil on the banks of the Solo River. There, a large group of convicts supervised by a couple of army corporals

shifted vast quantities of earth to uncover a hominid skullcap. This preserved pretty much the same parts as the disdained Neanderthal specimen,
but had housed a much smaller brain: some 940 milliliters (ml) in volume,
as opposed to 1,525 ml for the Neanderthal, about 1,350 ml for an average
modern human, and around 400 ml for a chimpanzee. The front of the
braincase resembled the Neanderthal in bearing a ridge above the eye sockets, but its form was different, and the cranial vault was even flatter. The age
of the fossil was uncertain, but there was no question that it was older than
the Neanderthals, and probably younger than the Himalayan apes.

What to make of this unprecedented find? At first, Dubois concluded
that his fossil was rather chimpanzee-like; but in 1892, after his workers had
recovered a very humanlike thigh bone in the Trinil deposits, he changed
his tune. In 1894, in a long monograph rather obscurely published in Java,
he argued that his small-brained new form, which he called *Pithecanthropus
erectus* (upright "ape-man"), had walked on two feet. However, it could not
be classified as either a human or an ape. Instead, it stood between the two,
in a succession that led from a gibbon-like ancestor, through the Himalayan
fossil forms, to modern humans (which from his point of view included
the Neanderthals). As a saltationist in the Huxleyan mold, he had no con-
ceptual problem with the gaps between these successive stages, or with the
conclusion that upright posture had preceded such distinctive modern human features as large brains and dexterous hands.

In broad outline, this was how Dubois was thinking when he returned
to Europe in 1895, bearing his fossil trophies. And although putting the
skullcap and femur into the same species was bound to be questioned (there
is actually still some uncertainty about the association, although there is no
doubt that the Javan form did indeed walk fully erect), many of Dubois' colleagues came to agree with him once they had seen the original specimens.
Among the broadly supporting voices was that of the German anatomist
Gustav Schwalbe, who rapidly produced influential monographs on both
Pithecanthropus and the Neanderthals. Feeling wounded—both by those
who rejected the association between skullcap and femur, and by what he
saw as Schwalbe's intellectual appropriation of his fossils—Dubois abandoned his own planned monograph on the specimens. He withdrew from
the fray, taking the fossils with him and thereby setting a pattern for the
future: paleoanthropologists have remarkably often treated access to the
fossils they control as a source of power or influence, or even as a means of

expressing disapproval, rather than as a vital way to provide the scientific community with the information it needs to refine its knowledge of human evolution.

As a result of Dubois' withdrawal, it was—inevitably—Schwalbe's secondary interpretation of those fossils that garnered most of the attention. And Schwalbe was pretty emphatic. While he considered the Trinil form closer to Neanderthals than to modern humans, he saw both hominids as components of a single evolving lineage, with *Pithecanthropus* ultimately giving rise to *Homo sapiens* via *H. neanderthalensis* (which he actually called *H. primigenius*). This is all understandable enough when you remember how tiny the human fossil record was at the time, and how anxious the early paleoanthropologists must have been to fit the little they knew within a logical and intelligible picture. But it is equally important to remember that it is in this rather impoverished context that the view became entrenched that *Pithecanthropus*, an odd fossil discovered in an outlying insular part of southern Asia, was a direct human progenitor.

THEORETICAL ADVANCES AND
MATERIAL EMBARRASSMENTS

It is often said that one of the weaknesses of Darwin's theory, as articulated in the *Origin*, is that he had no valid concept of how biological inheritance works. To a certain extent this is a valid criticism, since the whole idea of natural selection as a force for change depends on the passing of advantageous characteristics from one generation to the next. But at the same time, Darwin clearly did not need to know exactly how inheritance functioned to formulate his key concept of "descent with modification." Indeed, it was only after Darwin's death that the modern science of genetics was born. This was because, although the key principles of inheritance were actually articulated as early as 1866 by the Czech cleric Gregor Mendel, his findings languished in obscurity until they were independently rediscovered by several researchers at the turn of the twentieth century.

What Mendel had discovered was "particulate" inheritance. At the time he did his now-famous experiments on flowering peas in his monastery garden, most ideas of inheritance involved some kind of blending of parental characteristics. After all, offspring do tend to resemble both of their parents in one way or another. But Mendel realized that the actual units of inheritance—what we now call genes—do not blend. They may or may not be expressed in the appearance of the offspring—a dominant gene received from one parent might mask a recessive one received from the other—but they are nonetheless passed along intact in the egg and the sperm.

So far so good. But this does not explain how the novelties on which evolution depends are produced. This issue was finally solved in 1900, when the Dutch botanist Hugo de Vries not only rediscovered particulate inheritance but also exposed the phenomenon of mutation, in which, as we now know, copying errors are introduced into the genetic material when the reproductive cells are formed. Occurring spontaneously, and mostly producing a defective offspring or purely neutral results, occasionally a mutation will have a beneficial effect and provide the advantageous variations upon which Darwinian natural selection can go to work. In the early days, nobody had any idea what caused mutations, or how large their effects might typically be: de Vries himself speculated that mutations could produce distinctive new species in a single jump, supporting Huxley's view of evolutionary process. But experiments on fruit flies in the laboratory of the American biologist Thomas Hunt Morgan suggested otherwise, showing how single characteristics could be mutated while their possessors remained in the same species as their parents. Whoever might have been right, the science of genetics had arrived, and it would have a profound effect on orthodox evolutionary thought.

Meanwhile, paleoanthropology continued blithely on its own course, largely insulated from developments in evolutionary theory. In almost every case, new hominid fossils were discovered by practitioners of the nascent science of archaeology, who typically passed along their finds to human anatomists or other medical folk for study. These latter were often very acute observers, but they were not evolutionists who were likely to want to bring the study of human ancestry into line with the effort to understand biological diversity in the rest of the biological world. And ironically, the one case in which new remains came into the hands of a bona fide paleontologist yielded what is without question the single most embarrassing—and unfortunately best-known—chapter in the entire history of human evolutionary studies.

In 1908, workmen digging a pit at Piltdown in southern England handed some thick fragments of human skull to a local lawyer and antiquarian, Charles Dawson. Various other bits were discovered later, and in 1912 Dawson handed the lot to the distinguished fish paleontologist Arthur

Smith Woodward, Keeper of Geology at what was then called the British Museum (Natural History). Woodward initiated a formal excavation at the site, and soon much other material turned up, including stone tools and the fossil remains of extinct animals, some of which appeared older than those found in association with Neanderthals. Putting all the apparently human fragments together into a largely imaginary reconstructed skull (the original was missing its forehead, most of its right side, and, crucially, the points at which the cranium articulated with the jaw), Woodward pieced together a form that possessed a smallish though human-sized brain, together with an apelike jaw. Altogether, this form was pretty close to the reverse of what you would expect if the *Pithecanthropus* fossil were actually in the human line. But although the neuroanatomist Grafton Elliot Smith had pronounced the imprint of the brain on the inside of the skull bones to be "simian," most of the English establishment concurred with Woodward that here was a human ancestor of an age to rival the Java fossil. And the popular press went wild with the idea that the "missing link" had lived on English soil!

A second reconstruction of the Piltdown skull by the anatomist Arthur Keith increased the size of the brain, and even added a humanlike chin to the otherwise apelike jaw fragment. Keith himself thought the odd combination of braincase and jaw still indicated an evolutionary dead end rather than direct human ancestry; but he acknowledged that, should it be discovered that the canine tooth of the form Woodward had named *Eoanthropus* had been small like that of a human, rather than tall and pointy like that of an ape, the balance would be tipped in favor of human status. Voilà! A Piltdown canine was promptly discovered. Although apelike in other respects, this tooth was much lower-crowned than that of its ape counterparts and completed the picture of a human ancestor that had combined a large braincase with a reduced canine that had resided in an unfortunately apelike jaw. Keith duly concluded that Piltdown was at least marginally closer to modern humans than the Neanderthals were; as a result, by 1915 it was widely accepted that the Earliest Englishman had possessed a satisfyingly large brain, even if his jaw was rather apish. After all, an ancient ancestor surely had to be primitive in some respect.

The only problem was that, as revealed by chemical testing conducted some four decades later, the bones were not parts of the same individual. Nor even of the same species. The braincase from which the cranial fragments had been smashed was that of a modern human, while the craftily broken jaw had belonged to an orangutan. The canine also came from an ape, but it had been astutely filed down to remove its giveaway simian profile. The entire thing was a clever fraud involving the planting of entirely unassociated bones and artifacts at the Piltdown site. The pieces of "Eoanthropus" had been broken specifically to disguise their lack of association, and they were evidently intended to appeal to the mind-set favoring the view that, since modern humans are most strongly distinguished by their large brain, this feature must have been the defining feature of the entire human family. Whoever had planted the pieces knew his victims well.

The identity of the fraud's designer is still unknown for sure, and a whole cast of characters has been fingered, ranging from the Jesuit mystic and paleontologist Pierre Teilhard de Chardin (who, excavating alongside Woodward, had actually "discovered" the canine) to the novelist Arthur Conan Doyle, who had played golf nearby. All we can conclude with confidence is that Dawson was involved in some way. The motive for this elaborate scientific hoax is equally unclear, although it is evident that its author must have deeply disliked the English paleoanthropological establishment of the day. What is more, as the Piltdown bandwagon gathered momentum, the fraudster may actually have had second thoughts: eventually an artifact showed up at Piltdown that vaguely resembled a cricket bat—the very symbol of Englishness! The idea that the Earliest Englishman had possessed a large brain was at least a debatable one. But that he had played cricket should have beggared belief, as was probably intended. Yet even then, few in England suspected at the time that anything was wrong.

Outside Britain, attitudes were mixed. Across the Channel in France, the famous anatomist Marcellin Boule at first welcomed the Piltdown find. Only after a significant interlude of acceptance did French attitudes change to deep suspicion about the proper association of the cranium and the jaw. In America, though, the mammal specialist Gerrit S. Miller had already concluded by 1915 that the Piltdown find combined the cranium of a modern human with the mandible of a chimpanzee, and he made his viewpoint very clear.

With such diverse perspectives in play early on, it is clear in retrospect that the developing consensus that an ancient ancestor had given rise to a Piltdown-to–modern human lineage on the one hand, and to the poor benighted Neanderthals on the other, could never have lasted in the longer term. Accordingly, as time passed and one fossil find after another contradicted the "big brain first" theory, even British paleoanthropologists eventually began simply to ignore the Piltdown fossils. Still, a formal renouncement had to wait until 1953, when researchers found not only that the chemical makeup of the cranial and mandibular fragments of "Eoanthropus" indicated entirely separate origins, but also that microscopy revealed file marks on the crown of the canine tooth.

Many lessons may be learned from this unfortunate affair. The most obvious one is that, in the large and indispensable gray area of science—including much of paleoanthropology—in which propositions are not directly testable by experiment or observation, researchers can easily be led astray by their preconceptions. This is true even for the most meticulous of scientists, and it is hardly evident that in the early years of paleoanthropology its practitioners were particularly rigorous—or even that they were in a position to be, given the paucity of hard evidence. To make things worse, those years were the heyday of authoritative—and authoritarian—opinion in paleoanthropology, in which a scientist's position in the academic hierarchy often counted for a lot more than how well his or her beliefs were backed up by evidence. As my late great colleague Stephen Jay Gould observed, we are all unconscious victims of our preconceptions, and "the only palliations . . . are vigilance and scrutiny." Who could argue with that? And yet a recent study has concluded that even the hyperaware Gould was himself an unconscious victim of preconception when he excoriated the work of the nineteenth-century Philadelphia craniologist Samuel Morton for exactly the same fault. In criticizing Morton on what turned out to be dubious grounds, Gould unwittingly proved his own point, in the process underscoring its importance. Science is a self-correcting system of knowledge, but sometimes correction comes only slowly.

The Piltdown débâcle reminds us with unusual force that, when we look at any paleoanthropological—or other—question, we need always to examine our preconceived beliefs. But more than that, we need also to be aware of where those beliefs came from. Particularly if they were formed early enough in our experience, we may be unaware that they *are* preconceptions, rather than truths about the world that we are justified in taking for granted. We've seen that, from the very beginning, there have been alternative ways of construing both how evolutionary processes work and the nature of the species that result from them. And it turns out that what we are initially taught about such things has an enormous influence on how we will view them in the light of future evidence. To paleoanthropology's enormous detriment, this has brought with it a tendency to ignore Gould's wise exhortation as, come what may, practitioners try to fit new facts into existing frameworks (and, particularly, new fossils into existing species).

The second major lesson Piltdown teaches us is a practical one. Once a team of paleontologists has scoured a promising landscape for fossils eroding out of the rocks, the first task that faces it is to sort any fossils found into species groups. This is not always as easy as it might seem, and the problems are exacerbated when the fossils are fragmentary, as they usually are. The remains of dead animals rarely last long enough on a landscape to become covered by windblown or waterborne sediments and thus incorporated into the sedimentary rocks that build up in lake basins and on riverbanks. Even when bones do survive long enough for this to happen, they are likely to consist of widely flung bits and pieces that have been disarticulated and broken by predators and scavengers. What is more, once fossil remains are exposed at the surface once more by erosion of the rocks enclosing them, they are again at the mercy of the elements. The chances of their surviving long enough to be picked up by a passing paleontologist are slim indeed.

As a result of all these complex processes, complete fossil skeletons are a great rarity. Yet only when the bones of a single individual have survived together, preferably in lifelike articulation, is it possible to say with absolute confidence that all of those bones belonged to the same individual, and by extension to the same species. If bones of the same individual are actually found, it is much more likely that they will be in different places, sometimes quite far apart. Particularly if they are broken and missing vital parts, it may be very difficult to associate them with any certainty, and in general it's fair to say that any two bones found close together are more likely to belong to members of different species than they are to belong to the same individual.

The bottom line here is that associating bones found at the same locality with the same species, let alone the same individual, may often involve a leap of faith—as certainly it did at Piltdown. Yet this is not a lesson that paleoanthropologists over the years have seemed awfully anxious to take on board. Bolstered both by the gradualist notion of evolution inherited from Darwin, and by the material fact that there is indisputably only one species of hominid in the world today, the temptation is to reconstruct the history of humankind by projecting that one species back in time. This in turn produces a minimalist mind-set that wants to interpret all hominid fossils found at a particular site—or even in the same time zone, regardless of geography—as members of the same species. From there, it is not much of a leap to conclude that, because ancient hominids were thin on the landscape and thus relatively rare components of any fossil assemblage, all hominid fossils found in the same region are conspecific unless proven otherwise. As a result, it is not considered daring to assume that even hominid fossils found several miles apart in space, and hundreds of thousands of years apart in time, belonged to members of the same species—even where they consist of different parts of the body, for example a toe bone here, and a piece of mandible there.

Paleoanthropologists understandably tend to shudder when they hear the word "Piltdown." They would much prefer to forget this hugely embarrassing episode in their science. But they most emphatically shouldn't. Even if it remains largely unrecognized as such, Piltdown was a supremely teachable moment for paleoanthropology. Even today, a century later, we ignore its lessons at our peril.

CHAPTER 3

NEANDERTHALS
AND MAN-APES

AS THE INITIAL DEBATE OVER WHERE *PITHECANTHROPUS* FIT
into the picture of human evolution was still resonating both in the United
Kingdom and on the adjacent continent, the European hominid fossil record was expanding apace. Beginning in 1899, sediments filling the Croatian rock shelter of Krapina yielded an extensive, if very fragmentary, array
of Neanderthal bones. Between 1908 and 1911 several French sites—with
now magic names such as La Quina, Le Moustier, La Chapelle-aux-Saints,
and La Ferrassie—all yielded one or more Neanderthal skeletons of varying
completeness. In addition, in 1908 the German antiquarian Otto Schoetensack reported finding the lower jaw of a hominid in a quarry at Mauer,
near the city of Heidelberg. It didn't look very much like a Neanderthal, so
Schoetensack assigned it to the new species *Homo heidelbergensis*.

While all this was happening, geologists were doing their bit to clarify the broader chronological context of the European fossils. It had long
been known that, relatively recently in geological terms, the landscapes of
northern Europe had been affected by the presence of vast masses of ice
representing expansions of the Arctic ice cap. Toward the end of the nineteenth century, the Scottish geologist James Geikie proposed that these
"Ice Ages" had been organized into a series of cold "glacials," separated by

milder "interglacial" periods. In 1909 the German-Austrian geologists Albrecht Penck and Edouard Brückner formalized this insight into a sequence of four glacials (from earliest, the Günz, Mindel, Riss, and Würm, each named for a specific geographical feature with which it was associated). A fifth, the Donau, was later added at the beginning of the series. Between each glacial period, an interglacial signified the retreat of the ice cap. Penck and Brückner's signal achievement provided the first consistent temporal framework into which fossils could be integrated, and it involved sorting out some incredibly complex geology. The task was made all the more difficult by the fact that when a mass of ice spreads out over the landscape it tends to scour away, or at least to modify, the evidence left by its predecessors. What is more, to make life particularly tricky for archaeologists, melting ice caps release vast quantities of water that have an alarming tendency to wash out deposits from the caves and rock shelters in which they have accumulated. But despite the manifold practical difficulties in applying it, the Penck-Brückner sequence became the standard Ice Age time scale until it was edged out a half-century later by higher-tech approaches.

Geikie had already realized that some of the Paleolithic archaeological deposits known in Britain actually occurred within the Ice Age glacial sequence, and Otto Schoetensack believed that his Mauer mandible derived from early in the succession, namely from the Günz-Mindel interglacial. Of course, nobody knew at the time what that meant in terms of years, but it certainly implied that the Mauer mandible was particularly ancient: certainly older than the Neanderthal fossils, most of which appeared to come from considerably later in the sequence. In terms of the larger geological timescale, all of these glacial events and fossils lay within the Pleistocene epoch, named in 1839 by Charles Lyell to distinguish it from the preceding Pliocene in which rocks contained far fewer mollusks belonging to species that persist today. The Pleistocene is now broadly seen as the period within which the permanent Arctic ice cap has been in existence (about the last 2.5 million years); and since that ice cap is still around, albeit shrunken, there is an argument to be made that we are still in the Pleistocene epoch today. Nonetheless, the time since the final breakdown of the expanded European and North American ice sheets at the end of the last glacial, about 11,700 years ago, is conventionally included in a separate epoch, the Holocene. Moves are now afoot to recognize the period since humans began to alter the Earth's surface as the "Anthropocene," but many geologists disapprove.

With the formalization of a glacial sequence in Europe, rational discussion could begin of fossil hominids as players in a history that had roots in deep time and was characterized by a distinct sequence of events. The vanishing of the old diluvian/antediluvian argument meant that a chronology of human evolution could emerge—even if at this point it was only relative, and nobody had any idea just how much time had been involved. By the time World War I broke out, it had become evident that Neanderthals typically occurred alongside animals that preferred colder climates—although the Krapina hominids appeared to be an exception, perhaps belonging to the last interglacial. Neanderthal fossils had never been found in archaeological strata containing modern human remains, and although the first modern Europeans were also associated with Ice Age mammals, the Neanderthals appeared to be earlier in time. What's more, they always turned up in association with tools belonging to the earlier Mousterian culture, whereas modern humans seemed to have been responsible for the Aurignacian and later cultural expressions.

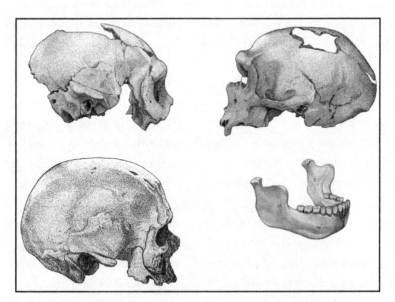

Four of the hominid fossils mentioned in this chapter. Above: crania of two Neanderthals, from Krapina, Croatia (left), and from La Chapelle-aux-Saints, France. Below left: the "Old Man" from the Cro-Magnon rock shelter, Les Eyzies-de-Tayac, France. Below right: the Mauer mandible from Germany, type specimen of Homo heidelbergensis. *To scale. Drawn by Don McGranaghan.*

In this way, it was credibly established by the early years of the twentieth century that Neanderthals had been the aboriginal inhabitants of middle Pleistocene Europe, and that they had been joined in the region by early moderns (known as Cro-Magnons for the French site at which their remains were first found) only at a relatively late stage. And while Eugene Dubois continued to dismiss the Neanderthals as a significant evolutionary phenomenon, he did believe that his *Pithecanthropus* derived from the late Pliocene or the early Pleistocene. Dubois' rough date estimate was entirely appropriate for a form that was ancestral to both Neanderthals and Cro-Magnons, and it was soon substantiated by a large assemblage of fossil mammals (alas, not including any hominids) excavated by German researchers at Trinil in 1907–08. In terms of chronology, at least, Gustav Schwalbe's linear scheme of 1899 would have appeared entirely vindicated, had it not been for the efforts of Marcellin Boule.

Boule, an accomplished anatomist, was the most influential French paleoanthropologist of his day. And it was to him that the fabulous new Neanderthal skeletons discovered at La Chapelle-aux-Saints, La Ferrassie, and other French sites were sent to be described and analyzed. While his counterparts in England were only too pleased to hail the Earliest Englishman from Piltdown as the ancestor of all humankind, Boule remarkably enough showed no such chauvinism in the case of the new French material. The key exemplar of the French Neanderthals was the amazingly complete skeleton from La Chapelle. This fossil was excavated in 1908 by amateur archaeologists who reported recovering it from a shallow pit dug into the bedrock that underlay a pile of stratified archaeological cave deposits. Argument still continues over whether or not the individual, a male old enough to have lost almost all of his teeth, had been deliberately, if simply, buried. But the mammal bones found in the strata above the skeleton belonged to species typical of cold conditions, suggesting that this Neanderthal had lived at some time in the last glacial. Recent dating of some of the mammal teeth to around 50,000 years confirms this age.

Between 1911 and 1913, Boule published an immensely influential three-part monograph on this hominid, in which he rejected the idea that it might have represented a human ancestor. As far as Boule was concerned, the Neanderthals such as La Chapelle were an evolutionary dead-end, living at a time when precursors of modern humans already existed—a notion that became known as the "presapiens hypothesis." He pointed out that the Mousterian industry of the Neanderthals had been replaced abruptly, everywhere, by industries associated with the Cro-Magnons, whose technologies were so sophisticated that he felt they must have been brought in from some other region of the world in which they had been developing for some time. In addition, Boule's examination of the La Chapelle skeleton indicated to him that its owner had possessed a divergent big toe, indicating a grasping foot with the weight-bearing characteristics of ape feet. His posture was hunched over; his knees were habitually bent; his neck was short and thick, his head jutting forward; and the contours of his large brain indicated intellectual inferiority. The very picture of a loser! What a contrast with the Cro-Magnons, whose "more elegant bodies . . . finer heads, large and upright foreheads . . . manual dexterity . . . inventive spirit . . . artistic and religious sensibilities . . . [and] capacities for abstract thought" made them "the first to deserve the glorious title of *Homo sapiens!*"

Although many aspects of Boule's anatomical observations on the La Chapelle Neanderthal have subsequently been cast into doubt, and despite the purple tincture of his prose, some parts of his description of *Homo neanderthalensis* retain considerable substance: perhaps one of the reasons why it was so influential at the time. More difficult to understand is Boule's treatment, or more accurately nontreatment, of the much-discussed *Pithecanthropus.* To place the Neanderthals in context, he embarked on a lengthy tour of the primate fossil record, discussing at length even such spectacularly irrelevant fossils as the extinct lemurs from Madagascar. But he paid barely any attention to the fossil from Java, dismissing it as a giant gibbon exactly as the perennially irascible Rudolf Virchow had done just after Dubois brought his find back to Europe. Even *H. heidelbergensis,* the oldest European human fossil then known, was relegated to the lowly position of Neanderthal ancestor. For the moment, the Piltdown "Eoanthropus" was Boule's preferred candidate for the ancestor of modern humanity, with all of its profound implications for paleoanthropology and for the interpretation of the Neanderthals.

And there matters rested for a decade or so, until the Hungarian American anthropologist Aleš Hrdlička, long deeply suspicious of the Piltdown remains, lectured London's Royal Anthropological Institute about "The Neanderthal Phase of Man" in 1927. The Neanderthals, Hrdlička declared, lay directly in the line of human descent; and, rather than being replaced by invading Aurignacians, they had evolved into them. According to Hrdlička, the variation seen among the rapidly accumulating Neanderthal fossils represented adaptive modification with advancing time—modification that was still continuing, for example in the reduction of tooth size. Similarly, the Mousterian Middle Paleolithic stone tool industry had evolved into the Upper Paleolithic industries of the Cro-Magnons. Perhaps largely because Piltdown still held center stage, few at the time took much notice of this reversion to Schwalbe's scheme, although in 1921 the linear interpretation was indirectly supported by Arthur Smith Woodward's interpretation of a cranium found in a mine at Broken Hill, in what is now Zambia. The age of this well-preserved skull was unknown, allowing Smith Woodward to speculate that *Homo rhodesiensis,* as he named it, was a potential intermediate between Neanderthals and *H. sapiens.* But before it could make much impact, the Zambian fossil was upstaged by the discovery of a truly ancient human precursor.

OUT OF AFRICA

In early 1925 Raymond Dart, a young Australian neuroanatomist who headed the department of anatomy at South Africa's University of the Witwatersrand Medical School, announced the serendipitous discovery of a curious new fossil in a lime mine at Taung, a couple of hundred miles to the southwest of Johannesburg. The title of Dart's paper said it all: *"Australopithecus africanus:* The Ape-Man of South Africa." Consisting of the face of a very young individual (whose first molar teeth had just erupted, as in modern six-year-olds) that was attached to a natural cast of its brain, this skull was unlike anything anyone had ever seen before. Clearly, though, it had belonged either to an ape with some humanlike features or to a human relative with some apelike characteristics.

Dart came down firmly in favor of "humanoid" status for his fossil, although many of the features that supported this interpretation were due at least in part to the individual's tender age: infant human and ape skulls resemble each other much more closely than adult skulls do. At 440 ml in volume, the fossil's brain was small, barely bigger than that of an ape of the same age; but Dart thought he could detect an expansion of the "higher centers" presaging the human condition. Indeed, he went so far as to surmise that his *Australopithecus africanus* had already "laid down the foundations of that discriminative knowledge of the appearance, feeling and sounds of things that was a necessary milestone in the acquisition of articulate speech." In Dart's estimation this conclusion was supported by his belief that the individual had lived in a semidesert environment that was hardly "favourable to higher primate life," and in which a hominoid's existence was made possible only by "enhanced cerebral powers." Human affinities were also implied by the placement beneath the skull of foramen magnum, the large orifice through which the brain connects to the spinal cord. In a quadrupedal chimpanzee this opening lies toward the back of the skull, whereas its forward position in the fossil suggested that the spine was carried upright—a quintessentially human trait.

When he presented it to the world, Dart had no idea what the age of his manlike ape was. Clearly it was old, but just how old he wasn't prepared even to speculate, because it was lacking geological context. The lime deposit at Taung was part of an ancient cave system that had formed in the water-soluble and ancient dolomitic limestones of the high veldt. At some time in the past, water had washed rubble from the surrounding landscape into an underground cavern, where rocky piles had accumulated that included animal bones cemented together by redeposited lime, interspersed with flowstones composed of pure lime. It was the lime deposits that the miners were after, and they got to them by blasting away the rocky breccias. In the process, they released fossils like the infant's skull from their stony prisons. In Dart's day the main way of dating geological deposits was still by the fossils they contained, and the baboon skulls also found in the deposit weren't very helpful in this regard. In a poignant vignette of early hominid life, it now seems most likely that the Taung infant was the victim of huge eagles that had nested above the inlet to the ancient cave, and that they had snatched their prey from the bosom of its family at around 2.8 million years ago.

On its own, though, the specimen's curious anatomy was compelling enough for Dart to conclude that he had before him the remains of a creature that was advanced enough to possess "just those characters . . . which are to be anticipated in an extinct link between man and his simian ancestor." So what, exactly, were they? Dart's conclusion on this point was less than entirely helpful. "A creature with anthropoid brain capacity, and lacking the . . . temporal expansions [of the brain] . . . necessary to articulate man," he declared, "is no true man. It is thus logically regarded as a man-like ape." For, "unlike Pithecanthropus . . . a caricature of precocious hominid failure," it could not be called an "ape-like man"—as this, for him, would have implied some degree of degeneracy. As far as Dart was concerned, it was far better for his fossil to be a manlike ape, striving toward perfection. In retrospect, the subsequent "man-ape or ape-man?" debate to which Dart's choice of terminology gave new life appears as one of the most unnecessary distractions in early-twentieth-century paleoanthropology; it took a long time to sink in that both man and ape are terms that specifically refer only to modern organisms, and that even fossils related to one or the other could be neither, without losing any significant aspects of their identities.

Crammed with imagination and wonderful phraseology, Dart's paper still makes a great read. But with paleontology thrust upon him by happenstance rather than through any design of his own, he was understandably thinking like an anatomist, rather than like a systematist interested in sorting out the wonderful diversity of Nature. He was happier trying to understand the implications of individual anatomical features than he was attempting to figure out where the whole organism fit into the greater scheme of things. Revolutionary as his find might have been, and courageous and perceptive in many ways as his analysis of it certainly was, he was still operating within the insular human anatomical worldview.

PEKING MAN

Nonetheless, a shared background didn't help Dart much with the anatomists who dominated the paleoanthropological establishment back in Britain. One or two of these colleagues were intrigued, but most—among them Arthur Keith, Grafton Elliot Smith, and Arthur Smith Woodward, all of Piltdown fame—airily dismissed the new fossil as an ape of one kind or another. And before Dart had his first opportunity to show it directly to his European colleagues in 1931, their attention—and that of the general public—had been distracted yet again. Between 1929 and 1934, Chinese excavators at a cave-fill site on the outskirts of Beijing (then known as Peking) recovered no fewer than 14 partial fossil hominid skulls, plus numerous other cranial and postcranial (body) fragments. Although it was obvious right away that the Peking fossils didn't differ hugely from Dubois' *Pithecanthropus* from Trinil, Davidson Black, the Canadian anatomist in charge of the excavations, gave them their own name: *Sinanthropus pekinensis*

(Chinese man of Peking). Braincase volumes came in at between 850 and 1,200 ml, all but one individual having a value greater than Trinil's. Along with a somewhat higher forehead, the slightly larger brain indicated to Black that at Peking he had a more "advanced" hominid. What's more, the Chinese form had one big advantage that the Javan one didn't: it had an archaeological context. In 1931, Black reported that some of the animal bones found at the cave alongside the hominids had been blackened by charring, indicating that, like some Neanderthals, *Sinanthropus* had mastered the domestic use of fire. At around the same time some simple stone tools were also recovered, and the combination provided Peking Man with the rudiments of a satisfyingly humanlike portrait.

In 1934 Black unexpectedly died, to be replaced by another human anatomist, the German Franz Weidenreich. This meticulous scientist, already renowned for his studies of Neanderthal-related fossils back in his now inhospitable homeland, exhaustively documented the collection of hominid remains from the cave now known as Zhoukoudian. In doing so he also rather ghoulishly amplified the emerging behavioral picture of Peking Man, based on his observation that the human remains from Zhoukoudian were extensively broken. The remains of some 40 hominid individuals, 15 of them children, had been excavated at the site, yet there was not a single complete skeleton among them. Indeed, there was hardly a complete bone. Why? Simple. However admirable its other qualities might have been, *Sinanthropus pekinensis* had been a cannibal. The broken bones were those of hapless hominids killed and eaten by their fellows, who had shattered their victims' bodies in the quest to absorb their spiritual and material qualities. Especially prized had been the brains, extracted through the smashed-away bases of the skulls. Half a world away, Marcellin Boule predictably disagreed. He felt that *Sinanthropus* was altogether too primitive to have made the tools found at the cave, or to have lit the fires. Harking back to his "presapiens" notion, he felt it much more likely that the broken remains had been feasted upon by members of a more advanced human species that had not left its remains at the cave.

In the long run it turned out that neither Weidenreich nor Boule had been right about this. Much more likely, the Zhoukoudian cave had been a carnivore den, to which hyenas had dragged parts of the unfortunate *Sinanthropus* individuals before crunching on their bones. The deposits Weidenreich supposed had been blackened by fire proved, much later, to have been stained by manganese. What is more, despite the curious fact that his viewpoint later became canonical in certain circles, Weidenreich was far from accurate in how he chose ultimately to integrate his *Sinanthropus* into the grander scheme of things. Immediately, though, his work had great impact. In 1937, Weidenreich visited the Dutch paleontologist Ralph von Koenigswald, who at the time was recovering additional *Pithecanthropus* material in Java's Sangiran basin, not too far from Trinil. The two correctly decided that *Sinanthropus* and *Pithecanthropus* were close relatives, although the Javan form was the earlier and more primitive. Not much more than a year later, the Japanese invasion forced Weidenreich to leave China for New York City (more successfully than the original *Sinanthropus* fossils, which were lost in the attempt to evacuate them). Once in New York he found himself with the luxury of time for reflection, and began to develop a theory of how the Javan and Chinese material fit into the overall picture of human evolution.

Weidenreich's major concern was how to account for the diversity of geographical variants of humankind in the world today. For there is no doubt that, on average, native people from Africa, Europe, Asia, and Australia exhibit distinctive aspects (however difficult it might be to draw sharp lines between these groups when push comes to shove—something less readily appreciated in Weidenreich's time than today). Still, at the same time people from all over the planet readily interbreed, as members of the same species invariably do. Weidenreich's response to this apparent paradox of anatomical diversity versus genetic integration was breathtakingly bold, and it would have flabbergasted an audience of evolutionary biologists interested in any other group of organisms. In 1947, in one of his last papers, Weidenreich declared his belief that "all primate forms recognized as hominids—no matter whether they lived in the past or live today—represent morphologically a unity when compared with other primate forms, and they can be regarded as *one species*" (his emphasis). Somehow, he had contrived to envisage a single human species that embraced the entire vast morphological variety that was known within the hominid fossil record, all the way from the bizarre Chinese form known as *Gigantopithecus* (actually, a fearsomely large fossil ape) to the slender, large-brained, and bipedal *Homo sapiens*. It was as if Weidenreich had envisaged all cats, lynx to lion, as one species, simply because they were all distinct from dogs!

Weidenreich summarized his view of human evolution in a rigorously symmetrical grid diagram, on which vertical lines represented different human lineages that had been in existence since "the very appearance of true hominids." Horizontal lines represented time, and each intersection point with a vertical line bore the name of a different kind of ancient or modern human—or would have, had our knowledge of our past been more complete. Finally, parallel diagonals symbolized gene flow among the various lineages, keeping them all united in one big happy species. In this way, for example, *Gigantopithecus* had given rise to *Sinanthropus*, which was the progenitor of the "Mongolian group" of modern humans. In a parallel sequence, *Meganthropus* (one of Koenigswald's Javan finds) yielded to the later *Pithecanthropus*, and ultimately to today's "Australian group." Various other names that had been applied to fossils found at various African and Eurasian sites were dotted around the grid, and the rigid directionality it showed was ascribed to some kind of inner urge among hominids that was linked to the enlargement of the human brain.

Weidenreich must be commended for his explicit articulation of what he believed to be an underlying pattern of human evolution, even as other experts continued to expound their seat-of-the-pants conclusions. But it remains true that his was a view that only an anatomist could have adopted. It showed a fine disdain—or at least disregard—for all of the rules and principles that over the years had been developed to govern systematics: the study of how the Earth's myriad species of organisms can be recognized, and how they are related to one another. On a more generalized level, it reflected a view of the evolutionary process—known as orthogenesis, the idea that evolution within lineages had some kind of innate direction—that, even before the 1940s, had begun to look distinctly outmoded.

Ironically, Weidenreich composed his final speculations about human evolution while he was working in New York City at the American Museum of Natural History, the institution that welcomed him following his flight from China. During the 1930s and 1940s, a major rethinking of the evolutionary process had already begun at and around the American Museum: a rethinking that was to make Weidenreich's beliefs on the matter appear hopelessly naïve. Poignantly, Weidenreich was smart enough to be fully aware of this. At one time he shared an office at the American Museum with the young paleontologist Bobb Schaeffer. As Bobb recounted it to me many years later, one day not long before his death at the age of 75, Weidenreich turned to him in the office and said, "I know that you young people have some new and really exciting ideas about evolution. But I hope you'll understand that I am just too old to change my ways of thinking."

THE NEW SYNTHESIS

So what were those new ideas that the aged Weidenreich was understandably reluctant to adopt (or even apparently to inform himself about, since this was the last time he ever raised the issue with Schaeffer, and they certainly did not feature in anything he wrote)? Well, they were actually the result of an extended process that unfolded over several decades. The rediscovery of Mendelian genetics at the turn of the twentieth century had spurred an enormous flurry of activity in the realm of evolutionary theory. As we've seen, Darwin himself had not needed to know precisely how heredity worked in order to come up with his succinct characterization of evolution as "descent with modification"; indeed, Darwin's own notion of how heredity might operate was way off the mark. But once the concept of particulate inheritance was in the air, and particularly once the idea of mutation actively began to be refined, many biologists perceived that here might be the key to explaining just what it was that drove evolutionary change. At first, almost every possibility was explored, from the putative role of "mutation pressure" (the rate at which mutations occurred) in propelling evolutionary trends, to that of the "hopeful monster" (spontaneous major genetic reorganizations) in generating new species. At first, few of the many theories of evolutionary change on offer were Darwinian in the sense that gradual natural selection was their key ingredient. But slowly, mathematical modeling of the behavior of genes in populations over time came to dominate, and quantitative geneticists such as Sewall Wright in America and Ronald Fisher in England began to develop an understanding of how the gene pools of organisms might respond to external selective influences.

This new approach was in turn made possible by a developing understanding of what genes themselves were. By early in the twentieth century, geneticists had already begun to envision genes as bead-like entities strung out along the chromosomes, those filament-like structures that are visible in the nuclei of cells about to divide. Each gene existed at its own locus along the chromosome, and came in alternative versions, known as alleles. The pairing of the chromosomes (one from each parent) explained the phenomena of dominance and recessiveness that Mendel had noticed: a dominant allele at a particular locus would mask the effects of its recessive twin, so that for recessive phenotypes (bodily structures) to be exhibited, identical alleles were needed from both parents. It also rapidly became evident that most phenotypic characteristics were controlled by more than one gene, while each gene was more than likely involved in influencing more than one character.

Particularly compelling among the emerging notions of how the behavior of genes in populations might influence evolutionary process was Sewall Wright's metaphor of the "adaptive landscape." The gene pool of any population contains many thousands of genes, and correspondingly more alleles; Wright reckoned that, in any given environmental situation, some combinations of alleles were likely to be more advantageous than others, enabling their "fitter" possessors to survive and reproduce more effectively than those less well-endowed. Accordingly, he created mathematical models, analogous to topographic maps, in which the fitter genotypes were clustered on the peaks, leaving the unfit to occupy the less-welcoming valleys. On this analogy, the problem facing any species was to maximize the proportion of the population occupying the hilltops, and to reduce the valley populations as much as possible. The key to accomplishing this was natural selection. Many other population geneticists latched on to Wright's formulation, often pushing it far beyond its author's original intent. But its major effect was to firmly associate population gene frequencies and natural selection, an association from which emerged what became known in the 1930s as the New Evolutionary Synthesis.

Many scientists contributed to the development of the Synthesis, but one name stands out: Theodosius Dobzhansky, whose 1937 masterpiece *Genetics and the Origin of Species* was the first comprehensive formulation of its principles. Dobzhansky, an old-fashioned naturalist as well as a gifted and innovative geneticist, worked at Columbia University, just uptown from the American Museum in New York City. And although he was deeply aware of the essential role of species in Nature, as of the role in evolution of genetic drift (a fancy name for random sampling error), he was in no doubt about the primacy of natural selection in shaping the history of life on this planet. First and foremost, evolution was a slow, gradual, long-term process that basically consisted of the accumulation within reproducing lineages of organisms of small genetic mutations, and of new combinations among existing alleles. The resulting incremental modification from one generation to the next played out via the inexorable working of natural selection: a blind process not directed toward any specific outcome but simply bound to happen as, out there on the ecological stage, fitter individuals outreproduced those of lesser natural endowment.

As environments changed, so would the pressures that assured the better-adapted would more effectively propagate themselves and their genes. Evolution was opportunistic, working ceaselessly to keep species optimized to their habitats. On this view, the modification of lineages was close to inevitable, either because environments were changing or because competition among individuals led to better adaptation to existing environments. Significantly, this within-lineage phenomenon could be extrapolated, via the splitting of lineages, to explain higher-level evolutionary phenomena such as the origin of new species, and of major groups such as whales, bats, and primates.

In 1942 the American Museum ornithologist Ernst Mayr contributed another great book, *Systematics and the Origin of Species*, which complemented Dobzhansky's work from the point of view of a systematist. Mayr pointed out that Darwin hadn't really dealt with the "origin of species" in his great work of 1859. To compensate for this deficiency, Mayr emphasized the importance of the natural discontinuities that species and "higher taxa" (families, orders, and so forth) represented, and elaborated the notion of "geographic speciation," whereby new species come into existence when an old one becomes divided by external events, as for example when a river changes course, or rising sea levels isolate populations of an old species on an island. Isolation thus became the key to innovation at the species level, particularly for complex vertebrates such as mammals; but ultimately natural selection drove the whole thing.

Two years later, the American Museum paleontologist George Gaylord Simpson brought his discipline into the fold with his book *Tempo and Mode in Evolution*. Simpson argued that Wright's adaptive landscapes were "like a choppy sea," which meant that Nature had to work constantly to keep organisms nicely balanced atop the peaks that were shifting under them. What's more, occasionally a peak would divide, carrying different parts of a population in different directions—and new lineages and new species would result. The upshot was that to Simpson, as to Darwin and Lamarck before him, lineages of organisms were dynamic things that transformed themselves out of existence. Species might have an existence in space, but over time they are ephemeral, thanks to the inexorable workings of omnipresent natural selection.

This view obviously holds practical difficulties for any paleontologist; for if species are constantly changing, how are we to recognize and delimit them among the fossils, which are all we have at our disposal to document the past? But the seductive reductionisms of the New Evolutionary Synthesis were so attractive that nobody at the time was bothered by such purely procedural problems. Indeed, they gave paleontologists reason to welcome the otherwise lamentable incompleteness of the fossil record, since it arbitrarily provided gaps at which species boundaries could conveniently be recognized. So any practical difficulties encountered by paleontologists in squeezing the known fossil record into the confines of the Synthesis tended to be brushed aside. And through the tireless efforts of Dobzhansky, Mayr, Simpson, and many others, by the late 1940s the Synthesis had come to dominate all areas of evolutionary biology in the English-speaking world—except for paleoanthropology. In that insular science the dominant human anatomists carried on their business of authoritarian pronouncement as usual, and largely in a theoretical vacuum. Weidenreich's archaic and far from universally admired scheme was the closest thing the discipline had at the time to a notion of process.

BACK TO THE FIELD

Fortunately, none of this impeded a steady drumbeat of hominid fossil discovery. In the decade following the rather equivocal reception of his *Australopithecus africanus,* Raymond Dart had done nothing to follow up on the original find. But in 1936 the physician and paleontologist Robert Broom announced the recovery of a rather battered adult skull of the same general kind from another limeworks site, this time more conveniently at Sterkfontein, a site much closer to Johannesburg than Taung was. Believing the associated fauna indicated his fossil to be a bit younger than Dart's, Broom initially called his new fossil *A. transvaalensis,* to reflect the South African province in which it had been found. Soon he changed his mind and transferred it to a new genus, *Plesianthropus* (near-man). But despite the difference in names, Broom had been convinced from the beginning that he now held in his hands full confirmation of Dart's claims. Redoubling his efforts to obtain better evidence, in 1938 he came up with a partial skull from another site at Kromdraai, just across the valley from Sterkfontein. This specimen differed from the earlier finds in having huge chewing teeth in a massively built face, leading him to dub it *Paranthropus robustus* (robust next-to-man).

Broom's Kromdraai discovery coincided with a visit to South Africa by the American Museum paleontologist William King Gregory and his dentist colleague Milo Hellman. These neutral observers were in no doubt about what Broom and Dart had found, and announced at a local meeting that the fossils were "in both a structural and a genetic sense the conservative cousins of the contemporary human branch." In other words, they were ancient human ancestors, somehow intermediate between the fossil apes known from the Miocene (the epoch preceding the Pliocene) and modern people. In a paper published in 1939, Gregory and Hellman formally placed them in the human family Hominidae: a fitting backdrop for the conclusions Broom presented in a monograph on his fossils (which by then included some postcranial bones as well as skull parts and teeth) published in 1946. Broom stated confidently that his australopiths (as they are now often known) had walked upright, and that although their faces and the teeth they contained were large, their dentitions were basically of human type. Most importantly, they had those reduced canine teeth that the forger of Piltdown had worked so hard to emulate. Their brains were small, but they probably had possessed basically humanlike hands that might even have made tools. His analysis of the animal bones suggested to him that the australopiths had lived in an open-country environment during the Pliocene epoch, an estimate more or less confirmed by modern dating that places the Kromdraai "robust" australopiths at a little under 2 million years old, and the more lightly built "gracile" Sterkfontein hominids at rather more than that.

Once again, the British panjandrums of paleoanthropology were unimpressed, at least until the rising star Wilfrid Le Gros Clark, professor of anatomy at Oxford, visited South Africa in 1947. Skeptical at first, Clark was rapidly convinced by Broom's claims. On his return to England, he successfully converted his colleagues—with the notable exception of the South African expatriate anatomist Solly Zuckerman—to the view that the australopiths really were early hominids. The exception was an important one, though, because Zuckerman, who had reportedly employed his anatomical

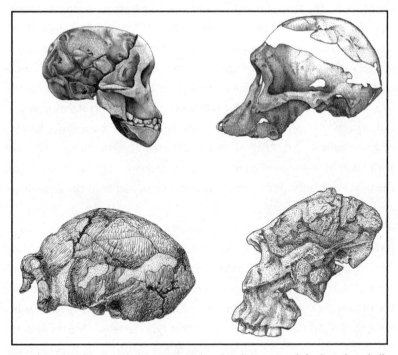

Four more of the hominid fossils mentioned in this chapter. Top left: the infant skull from Taung, South Africa, type specimen of Australopithecus africanus. *Top right: an adult cranium of* A. africanus, *from Sterkfontein, South Africa. Bottom left: "Skull XII"* Homo erectus *from Choukoutien, China. Bottom right: Partial cranium of* Paranthropus robustus, *from Swartkrans, South Africa. To scale. All drawn by Don McGranaghan except bottom left, by Diana Salles.*

expertise to design bombs that blew people apart more effectively, was for many years an influential scientific adviser to the British government. His baleful influence still resonated in paleoanthropology right through the 1960s, when I was first becoming involved in the field.

Clark himself was an interesting figure, and he was one of the few anatomists interested in the human fossil record who had bothered to think deeply not just about what paleoanthropologically inclined anatomists were doing, but about how they were doing it. He fully realized that, in taking on the fossils, human anatomists were assuming a duty to properly characterize the species they were dealing with, and to understand what they were doing when figuring out the relationships among them. Still, his prescription for carrying out this responsibility was a typical anatomist's solution to the problem: the analyst should not be swayed by any individual feature of a particular fossil, but instead should look at its "total morphological pattern." It all sounded very sensible, but the problem lay—and unfortunately still lies—in knowing just what a total morphological pattern is. There is nothing in this concept that is susceptible of being characterized, or quantified, or compared directly with anything else. In practice, total morphological pattern just boils down to the impression that the fossil you're looking at makes on you. And that, of course, left each expert with nothing more than his or her own personal reactions, conveniently expressed through the prerogative of authoritative pronouncement that had been the anatomists' *modus operandi* all along.

Of course, by the late 1940s subjective judgment had undeniably brought paleoanthropology—and paleontology in general—a very long way; and further fossils found at Sterkfontein rapidly confirmed Broom's opinion that here was a kind of "ancestral man just a little more primitive than *Pithecanthropus*." What's more, the nearby site of Swartkrans was beginning to yield more fossils of the robust type, to which Broom gave the name of *Paranthropus crassidens* in celebration of their large chewing teeth. By 1949 Swartkrans had also produced a couple of jaws of a more lightly built hominid. Broom and his associate John Robinson baptized these *Telanthropus capensis* (far man from the Cape) and suggested their new genus might be "intermediate between one of the ape-men and true man," hence putatively ancestral to *Homo heidelbergensis* in Europe.

To add to the wealth of new hominid fossils and names that were now almost literally spewing forth from the South African limestone caves around Johannesburg, Dart himself was developing an interest in fossils reported from a lime mine at Makapansgat, well to the north of the city. Upon close inspection in 1946 and 1947, the Makapansgat miners' dumps proved to be full of rather fragmentary fossils, including a few hominids of the more lightly built sort. Dart gave these latter the name *Australopithecus prometheus*, in the mistaken belief that the blackened condition of many of the bones indicated charring by fire. And he went further. The broken bones—of australopiths and other mammals alike—recalled to him the fragmented skulls from Zhoukoudian that Weidenreich had believed to be the remnants of cannibal feasts, and Dart let his imagination run wild. He concluded that the fractured fossils must have accumulated as a result of the slaughtering, butchering, and cooking activities of australopiths, who had possessed an "osteodontokeratic" (bone, tooth, horn) material culture and had used these materials as tools (bone hammers, dental saws, horn-core daggers). The basics of modern human nature were already established among the australopiths, because "to that osteodontokeratic psychological indoctrination humanity added stone and metal, but has never been able to free itself . . . from bone, tooth and horn." Here was born the image, brought so vividly to the page by Robert Ardrey and to the screen by Stanley Kubrick, of ancestral hominids as bludgeon-wielding "murderers and flesh hunters" (as Dart put it), whose innate violence led inevitably to the "blood-spattered, slaughter-gutted archives of human history."

There is only one problem with this scenario: drama is basically its only virtue. On closer scrutiny the evidence for fire vanishes, the fossils turn out to have been broken by natural processes, and the dense accumulations of pieces of carcass at Makapansgat are most plausibly due to the activities of porcupines, which occasionally feed on carrion and bring bones back to their nests to gnaw upon. Indeed, far from being bloodthirsty predators, the small and biologically defenseless australopiths—remember, they had small canine teeth!—were almost certainly routinely the prey of large carnivores, perhaps particularly leopards. An immature australopith braincase from Swartkrans poignantly reminds us of this: it bears twin round holes into which the lower canines of a leopard perfectly fit.

OUTSIDE AFRICA

Still, while interpretations come and go the fossils remain. And thanks to those South African discoveries, by the middle of the twentieth century the human fossil record was beginning to take on the basic form we recognize today. Early bipeds with small brains and big faces had existed in Africa during the Pliocene; the taller and larger-brained Chinese and Javan forms fell somewhat later in time; and the big-brained Neanderthals—perhaps slightly preceded by crania found at Swanscombe in England and Stein-heim in Germany—had coexisted with early modern humans in Europe. But some additional complications had also become apparent at the recent end of the timescale. During the early 1930s the English archaeologist Dorothy Garrod exhumed a number of hominid fossils at Tabūn and Skhūl, two adjacent rock-shelter sites on the seaward slopes of Mount Carmel, in (then) Palestine. Both sites appeared to have been formed in the last interglacial period, and both yielded Levalloiso-Mousterian stone tools, basically equivalent to what the penecontemporaneous Neanderthals were manufacturing in Europe. But while a rather lightly built female skeleton and a more robust jaw from Tabūn looked Neanderthal-like, the several individuals from Skhūl, a site that appeared to be a veritable cemetery, were distinctly different. Their braincases were higher and rounder than those of Neanderthals; their browridge structure was distinctive; and their faces, though forwardly jutting, did not have the distinctively inflated midsection that was turning out to be typical for the Neanderthals.

The hominids from both sites were described together in 1939 by the Berkeley anthropologist Theodore McCown, in collaboration with the ubiquitous anatomist Sir Arthur Keith. It is pretty evident from reading their breathtakingly opaque conclusions that these two scientists did not see eye-to-eye on these fossils, but in the end they opted to view them all as members of the same highly variable species, to which they gave the name

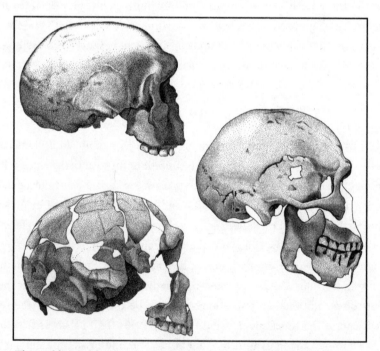

Three additional fossils mentioned in this chapter. Top left: cranium from Steinheim, Germany. Bottom left: Neanderthal cranium from Tabūn, Israel. Right: skull V from Skhūl, Israel. To scale. Drawn by Don McGranaghan.

of *Palaeoanthropus palestinensis*. Tying themselves into knots, they speculated that the totally implausible degree of variability seen in the supposed Mount Carmel population was due either to "the throes of an evolutionary transition" in which the hominids were "unstable and plastic in their genetic constitution," or to "hybridity, a mingling of two diverse peoples or races." These two possibilities have basically set the parameters of discussion of the Skhūl (and occasionally the Tabūn) hominids ever since, but the opacity of McCown and Keith's discussion underscores the perils of allowing oneself to be influenced by factors other than anatomy (in this case, by the similarity of the associated stone tool kits) when making judgments about what your fossils are. A fundamental lesson, one might think; but it is one still largely unlearned in paleoanthropology.

POPULATION THINKING

The high pace of discovery and description of new hominid fossils over the first half of the twentieth century, combined with the general lack of interest among paleoanthropologists in systematic norms, led to a huge multiplication of species and genus names. It was almost as if each new fossil assemblage that was found needed to be baptized with its own genus and species name (just as human individuals have their family and given names) if it were to be properly discussed. Indeed, by the time the late 1940s came around, at least 15 different genus names were being actively applied to one fossil hominid or another, giving the superficial impression of an astonishing diversity within the still relatively poorly known hominid family.

One scientist who was having none of this was Theodosius Dobzhansky. The New Evolutionary Synthesis, of which he was an intellectual father, had been born out of what his colleague Ernst Mayr dubbed "population thinking." Many early geneticists had thought of species as little more than convenient bundles of the characters whose inheritance they were interested in studying, while old-fashioned systematists had been prone to look upon species as "types" of organisms. In contrast to both, population thinkers recognized that species were actually clusters of individuals bound together by reproductive continuity, regardless of what they looked like to the human eye. In other words, as I've previously noted, individuals did not belong to the same species because they looked similar; they looked similar because they belonged to the same species. In addition, many species were in fact "polytypic," which is to say they were differentiated into recognizable geographical varieties that could potentially interbreed, but were prevented from doing so by topographic or environmental circumstances. Such local varieties—subspecies—were an engine of evolutionary change, supplying the raw material for new species.

Dobzhansky applied this kind of thinking to the human fossil record in 1944, and although it is doubtful that he had actually ever seen any original human fossils, he was prepared to make some pretty definitive assertions. One of these was that "the differences between Peking and Java men are easily within the magnitude range of the differences between the living human races." Another was that the variation seen at the Mount Carmel sites was due to hybridization between Neanderthals and modern humans, subspecies that had differentiated in different regions but had come into contact in Palestine. From there it was but a short step to conclude that "in Hominidae, a morphological gap as great as that between the Neanderthal and the modern may occur between races rather than species." Having established this, at least to his own satisfaction, Dobzhansky proceeded to contrast what he considered to be the two leading models of human evolution. One was the "classic" view (which emerged from that mindless plethora of genera), of a "tree with many branches . . . [of which] the known fossils represent . . . only rarely the main phylogenetic trunk." The other was Weidenreich's "parallel development of races." And what Dobzhansky concluded—unsurprisingly, given the amount of morphological variation he was prepared to accept within a single species—was that the differences between the two models were of little significance, because everything that had transpired in human evolution since the days of Java Man had occurred within the confines of one polytypic species. Presciently, Dobzhansky envisioned a hugely complex set of regional differentiations and hybridizations during the Pleistocene; but in brushing a huge range of morphologies under the single-species rug, he was setting the stage for a distorted view of the human fossil record that lingers tenaciously.

CHAPTER 4

THE SYNTHESIS AND HANDY MAN

IF I HAD TO OPT FOR ONE SINGLE YEAR AS THE MOST MOMEN-
tous in the twentieth-century intellectual history of paleoanthropology, I
would unhesitatingly choose 1950. Theodosius Dobzhansky had, of course,
already put the Synthesis cat among the paleoanthropological pigeons back
in 1944, but it was wartime, and nobody seems to have taken much immediate notice. Nonetheless, Dobzhansky's take on human evolution pointed
to the future, and in many ways the end of World War II, the year after his
article appeared, also marked the passing of the old guard in paleoanthropology. In 1948 the aged though still-industrious Arthur Keith published
a volume entitled *A New Theory of Human Evolution*, but the book actually did little to deliver on its title. It is mostly remembered, if at all, for its
vaguely anti-Semitic stance. The time had come for a new cast of characters
to step onto the paleoanthropological stage.

A leader among the new generation of biological anthropologists was
Sherwood Washburn. Rather conventionally trained at Harvard during the
1930s, Washburn enthusiastically embraced the New Evolutionary Synthesis after joining Dobzhansky on the Columbia faculty in 1940. And it
was with this energetic convert that the Synthesis at last acquired a conduit
into paleoanthropology. In 1950 Washburn (by then at the University of

Chicago) and Dobzhansky jointly organized a conference hosted by Long Island's Cold Spring Harbor Laboratory. Grandly titled "The Origin and Evolution of Man," this international meeting brought together numerous luminaries of paleoanthropology and adjacent sciences, including all three of the giants of the Synthesis. It was thus loaded with star power, but in retrospect, one contribution stands out not only as the most newsworthy presentation at the conference, but also as one of the most influential benchmarks ever in paleoanthropology. Significantly, it was not made by a paleoanthropologist. It was made by the ornithologist Ernst Mayr.

As forceful on the printed page as in oratory—although his published version bears all the marks of haste in preparation—Mayr didn't bother to mince his words. In no uncertain terms, he informed the assembled multitude that the picture of complexity in human evolution implied by all those hominid species and genera was just plain wrong. To begin with, he declared, both the theoretical and the morphological yardsticks by which the anatomists had differentiated them were entirely inappropriate. For example, if you took a couple of fruit fly species and blew them up to human size, they would look much more different from one another than the members of any pair of living primate species do. And the same went in spades for fossil hominids.

Spectacularly irrelevant as the metaphor was, it resonated with an audience that was uncomfortably aware of the thin theoretical ice on which it skated. And it primed that audience for Mayr's more specific claim, that the supposed diversity of hominid genera and species just didn't exist. What was more, Mayr continued, even in principle there was no way in which that diversity could have existed, because the possession of material culture so remarkably broadened the ecological niche of tool-wielding hominids that there would never have been enough ecological space in the world for more than one human species at a time.

Put together, Mayr said, these various practical and theoretical considerations dictated that every one of the human fossils known should be placed within a single evolving polytypic lineage. And not only was there a mere three species recognizable within that lineage, but every one of those species belonged to a single genus: *Homo*. As Mayr saw it, *Homo transvaalensis* (the australopiths) had given rise to *H. erectus* (including *Pithecanthropus*, *Sinanthropus*, and so forth), which in turn evolved into *H. sapiens* (including the Neanderthals). And that was it.

Still—as if he somehow felt that things couldn't have been quite this simple—Mayr inquired explicitly why, unlike virtually any other successful mammal family, Hominidae had not thrown off a whole array of species. "What," he asked, "is the cause of this puzzling trait of the hominid stock to stop speciating in spite of its eminent evolutionary success?" His ingenious answer to this excellent question brought him right back to "man's great ecological diversity." Humans, Mayr declared, had "specialized in despecialization." What was more, "Man occupies more ecological niches than any known animal. If the single species man occupies all the niches that are open for a *Homo*-like creature, it is obvious that he *cannot* speciate" (emphasis mine). Mayr also noted something else very special about "man," something that, in his view at least, supported his reconstruction of human phylogeny as an infinite recession of today's ubiquitous *Homo sapiens* back into the past: "Man is apparently particularly intolerant of competitors . . . the elimination of Neanderthal man by the invading Cro-Magnon man is only one example."

Mayr took questions at the end of his presentation. When asked (not by a paleoanthropologist, of course) about how the notable morphological differences found among fossil hominids could all be compressed into a single genus, he finessed his answer by responding that "since there are no absolute generic characters, it is impossible to define and delimit genera on a purely morphological basis." Nobody at the time saw fit to call him on this. Nobody pointed out the obvious: that morphology was the only thing that paleontologists had to work with, and that, while he might technically have been right about the nonexistence of "absolute generic characters"—whatever exactly that meant—fossil genera had to be recognized from their morphology. Nor did anybody suggest that intolerance of competition might be specifically a feature of *Homo sapiens*, distinguishing it from even its closest relatives. And neither did anyone question any other of Mayr's sweeping and hugely speculative declarations—either at the time, or in the couple of years following the appearance in print of his provocative comments.

Almost certainly, the reason for this supine acceptance of his many-sided criticisms of their field is that Mayr's broadside had shocked the tiny elite of paleoanthropologists into some long-overdue introspection. They finally began to realize that they and their predecessors had been operating in a theoretical vacuum, in which nobody—except perhaps Franz Weidenreich—had bothered to think much either about the processes that might have under-written the stories they were telling about their fossils, or about how their operating assumptions fit in with what was known about how the rest of Nature had evolved. And here was Mayr, the self-assured architect of the Synthesis, with an eloquent and comprehensive analysis of their science: an analysis that combined a nod to morphology with considerations of evolu-tionary process, systematics, speciation theory, and ecology—all those key factors that paleoanthropologists were now beginning to feel guilty about having largely ignored—to produce a cogent and coherent statement about human evolution. Without an intellectual fallback position, what could they do but capitulate? Caught in this uncomfortable epistemological situation, hardly anybody seemed to mind that Mayr's scenario was far from firmly anchored in the study of the fossils themselves.

The major English-speaking exception to this instant surrender was Robert Broom's younger associate John Robinson, who pointed out at some length that the morphological heterogeneity among the gracile and robust australopiths—and some similarities he saw between some South African and early Javan material—indicated at least two coexisting hominid lin-eages in the Pliocene or early Pleistocene. And although Mayr's grudging admission that Robinson indeed had a valid point was buried in a pile of notes published in a journal that paleoanthropologists didn't read, once Robinson had pointed this out most of his colleagues came to agree that the robust australopiths were best excluded from Mayr's linear scheme. The genus name *Australopithecus* continued to be used for all the gracile aus-tralopiths (and for some paleoanthropologists, mainly Robinson, the ro-bust offshoot continued to be called *Paranthropus*). Robinson himself also continued to use the name *"Telanthropus"* (his quotes) for the mysterious, very lightly built hominid fossils from Swartkrans—and by then also from one section of Sterkfontein, which by the mid-1950s also had begun to pro-duce some crudely flaked stone tools.

But that was about it. After that fateful year of 1950, paleoanthropolo-gists in the English-speaking world dutifully lined up behind Mayr's con-tention that, after the australopith stage (and probably well back into it), hominid evolution had to all intents and purposes consisted of the pro-gressive modification of a single central lineage. At any one point in time that lineage had consisted of multiple geographical variants, but the whole thing was consistently knit together by genetic interchange. Throughout, the central tendency was toward a slow, long-term evolutionary burnishing under natural selection—which had acted in subtly varying ways in different parts of the world, but never strongly enough to lead to speciation. Change of this kind could conveniently be tracked by following the evolution of individual features—the brain, say, or the foot, or the gut—without any perceived need to follow the entities—the species—into which those features were bundled.

Yet there was that pesky question of morphology. Even in the 1950s, there really was quite a bit of anatomical diversity out there in the known hominid fossil record: diversity that students of any other (normally speciating, non-"despecializing") group of mammals would have readily acknowledged with multiple species names. But paleoanthropologists had a ready response to this conundrum: they ignored it. Indeed, for a decade and more following Mayr's attack, they barely dared to use formal species names at all. The typical human phylogeny diagram of the period was a blobby outline circumscribing the hominid family, with geographic lobes filled with the names of individual fossils—Mauer, maybe, or La Chapelle, or Broken Hill—which had become the *de facto* units of discussion in the field. The evolutionary relationships of those individual fossils were loftily discussed, without any perceived need to worry about the species to which they had belonged.

Given the circumstances, this avoidance of formal zoological names and the species they denoted was perhaps understandable. What's more, it wasn't entirely a bad thing. Indeed, in retrospect it appears as a phase that paleoanthropology desperately needed to go through. As Mayr had so forcefully pointed out, the huge mass of formal zoological names that had accumulated for a relatively limited collection of fossils was totally out of proportion to reality. A clearing-out was urgently in order before a rational appreciation of diversity could be built up. And this is what actually happened, although the diversity part subsequently occurred only to a severely limited extent. Because although an unspoken consensus emerged that a scientist could at least eventually be forgiven for applying a new name to a really distinctive new fossil, it remained absolutely taboo to give a new name to a fossil that was already in the literature, or even to resurrect an old disused name for it.

Mayr's strictures did not just result in a salutary taxonomic house cleaning. They also opened the way to a much clearer framework for interpreting the hominid fossil record. Exemplary was an analysis of the Neanderthals published in 1951 by F. Clark Howell of the University of Chicago. By then the formal declaration that the Piltdown "fossil" was a fraud had not quite yet been issued by the British Museum (Natural History) in which the bones resided; but it was already widely known that chemical tests had proven the skull and mandible fragments to be unassociated, and Howell was operating in what was at last an officially Piltdown-free world. Imbued by the spirit of the Synthesis, Howell's paper on the Neanderthals envisaged a single European lineage that led from Mauer, through crania discovered at Swanscombe in England and Steinheim in Germany, to an "early Neanderthal" assemblage represented by crania found in quarries at Saccopastore in Italy and Ehringsdorf in Germany (the latter described by Weidenreich before he went to China). The early Neanderthals appeared to date from the last (Riss-Würm) interglacial, and had less emphatically Neanderthal features than the western European "classic" forms of the last glacial (among them La Chapelle, La Ferrassie and Feldhofer), to which they had given rise.

Howell also thought he could detect a geographical trend, from a lighter build in the east of Europe to the more pronounced features of the classics in the west. Putting all this together produced a scenario whereby early Neanderthals in eastern Europe and the Levant had given issue to modern *Homo sapiens* via the Mount Carmel forms. The newly minted moderns then spread westward, ultimately to eliminate their classic Neanderthal cousins.

All this evolutionary activity had been mediated by dramatic shifts in Ice Age climates. In the warmer interglacial conditions, Neanderthals were more or less continuously distributed across Europe. When cooling started at the beginning of the Würm glacial period, the western and eastern populations were cut off from one another by deteriorating conditions in the intermediate zone. This allowed them to start diverging. Harsh climates and strong natural selection (and maybe a bit of genetic drift) in the north and west subsequently resulted in the classic Neanderthal morphology, while in the kinder south selection was less stringent, leading to a reduced rate of change.

If it is permissible to divide up the history of paleoanthropology into intellectually distinct periods, Howell's thoughtful analysis of the Neanderthals can legitimately be seen as inaugurating paleoanthropology's modern epoch. To the contemporary ear there is an archaic ring to virtually every-thing written in the field before Mayr's onslaught. In contrast, infused as it was with the teachings of the Synthesis, Howell's contribution was a beacon of future perspective. In its pages morphology, time, geography, climate, and process (as specially tailored by Mayr for anthropology) were all incor-porated into a single coherent and global view of Neanderthal evolution, in a way that was entirely new and exciting. Humans love stories, and Howell told a rollicking good one. Still, in the decades that were to follow not all paleoanthropologists would prove to be as sensitive to matters of process as Howell clearly was.

RADIOCARBON DATING

As it happened, 1950 was also the year in which radiocarbon dating was invented by the University of Chicago physical chemist Willard F. Libby. For the first time, it became possible to assign ages in years to certain hu-man fossils, rather than simply placing them in sequence. Like most of the many other methods of "absolute" dating that have subsequently been developed, the radiocarbon technique depended on radioactivity, the phe-nomenon whereby certain naturally occurring unstable atoms spontane-ously decay to stable states at constant and measurable rates. The rate is expressed as the "half-life" of the particular kind of atom, that is, the time it takes for half of the atoms in any sample to decay. Radiocarbon dating depends on the decay of ^{14}C, an unstable form of carbon that is maintained during life as a constant proportion of the carbon present in all organisms. When the organism dies its ^{14}C ceases to be renewed, and begins to decline as a proportion of total carbon. So if you can determine the amount of ^{14}C present in a sample relative to the amount of stable carbon in it, you have a way of discovering how much time has elapsed since the death of the organism.

Radioactive carbon has a rather short half-life of 5,730 years. This caps the useful time range of the method at about 40,000 to 50,000 years (some-where in the latest Pleistocene), after which there simply isn't enough ^{14}C left to measure. What is more, in the early days, you had to destroy a good chunk of material to get a date, with the result that fossils themselves were rarely dated. Instead, dating was done on materials associated in the same archaeological deposits—charcoal was, and still is, a favorite. In recent years methods have become available that allow tiny samples to be dated, so that increasingly sophisticated radiocarbon techniques are routinely used directly on fossils themselves, or on such materials as the charcoal sometimes employed as the black pigment in cave paintings. Still, the caveat remains that organic material needs to be preserved in your sample—a condition that is not always satisfied.

The classic Paleolithic archaeological site at which the radiocarbon method was employed early on was the rock shelter of Abri Pataud. Situated in Les Eyzies, a little town in western France situated at the center of one of the most extraordinary concentrations of Paleolithic sites known anywhere, and containing a thick pile of sediments representing almost all the first half of the Upper Paleolithic, Abri Pataud was an ideal place to deploy the new dating technique. The Aurignacian layers at the bottom of the sediment pile turned out to date between about 30,000 and 34,000 years ago, while the late Solutrean/Protomagdalenian levels at the top came in at around 22,000 years (the most recent analyses place them a little earlier). Ultimately the Pataud rock shelter's excavator, the Harvard archaeologist Hallam L. Movius, was able to propose a timescale for the entire Paleolithic in western France that was later fine-tuned by others. In broad terms, the Upper Paleolithic in southwestern France began a little over 40,000 years ago as the first *Homo sapiens* began seeping into the region. The Aurignacian culture was replaced around 28,000 years ago by the Gravettian, which in turn yielded regionally to the Solutrean at about 22,000 years ago—although it persisted much longer farther east. The art-drenched Magdalenian culture replaced the Solutrean at around 18,000 years ago, persisting until around 10,000 to 11,000 years ago, when warming at the end of the last Ice Age saw expanding forests drive away the large herds of grazing animals that had sustained the hunting peoples of the Upper Paleolithic.

The Mousterian industries typical of Neanderthals in western France lie mostly beyond the range of radiocarbon. There are few convincing dates of under 40,000 years. Intriguingly, an additional "Châtelperronian" culture, represented in central and southwestern France and parts of northern Spain at sites between about 44,000 and 40,000 years old, has aspects that reflect both Mousterian and Aurignacian approaches to stone toolmaking. It has also, and very controversially, been associated with decorative objects. However, it is clear that in historical terms the Châtelperronian does not represent a biocultural intermediate between the Mousterian and the

Aurignacian. For one thing, in certain places full-blown Mousterian industries are found above it in the stratigraphic succession. For another, in those rare instances where human fossils are associated with the Châtelperronian, they are in all cases distinctively Neanderthal.

HANDY MAN

Into the new and rather humbled post-1950 paleoanthropological milieu strode Louis Leakey. In many ways Leakey was the ultimate outsider, and he was never one to yield the canonical authority conferred by the control of important fossils to effete intellectuals who fretted about such niceties as evolutionary process. Kenya-born to missionary parents in 1903, the charismatic, energetic, and opinionated Leakey was educated in archaeology and anthropology at Cambridge, where he honed his interests in the prehistory of East Africa without apparently ever acquiring much appreciation of the subtleties of systematics. Following a marital scandal that precluded his securing a respectable position in the United Kingdom, Leakey established himself as curator of the Coryndon Museum in Nairobi, precursor of today's National Museums of Kenya.

For decades, operating on a shoestring budget and largely in obscurity, Leakey and his new archaeologist wife Mary scoured the landscapes of East Africa for evidence of the early hominids and apes that they viscerally *knew* had lived there. Ultimately, they centered their activities at Olduvai Gorge in northern Tanganyika, not far from the famous Ngorongoro Crater and the Kenya border. Since the 1930s Leakey had known that ancient stone tools eroded in abundance from the walls of this great scar in the earth, and he dreamed of finding the fossil remains of the hominids that had manufactured these testaments to true humanity.

For years, the Leakeys' incredibly arduous searching yielded no more than the occasional hominid jaw fragment or tooth. But in 1959 all those unrewarded years of gritty determination finally paid off. The major rock units of the Gorge were exposed in its walls like the layers in a cake, and were identified as Beds I to IV, counting up from the bottom. While revisiting a locality right at the base of the sequence that, because of all the stone tools scattered around, she and her husband had for many years regarded as an early hominid "living site," Mary Leakey discovered most of a magnificently preserved hominid cranium (skull without mandible). This new

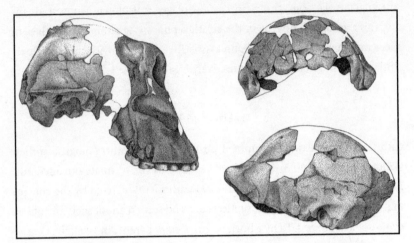

Three hominid crania from Olduvai Gorge. Left: Bed I OH 5, "Zinjanthropus," type specimen of Paranthropus boisei. *Top right: skullcap of lower Bed II* Homo habilis, *OH 13. Bottom right: upper Bed II OH 9, initially compared to* Homo erectus. *To scale. Drawn by Don McGranaghan.*

find turned out to be most comparable to the robust australopiths that John Robinson had been busily excavating at Swartkrans in South Africa, and which he continued to identify as *Paranthropus*. But it was even more massively built than they were, with tiny incisor and canine teeth lying in front of huge, flat premolars and molars that caused Louis to nickname the fossil "Nutcracker Man."

This new Olduvai fossil was an astonishing and unprecedented find that finally placed East Africa on the paleoanthropological map. Still, initial joy at the presumed discovery of the stone toolmaker of Olduvai masked a deeper level of disappointment, because the new find simply did not meet expectations. Leakey was a firm believer in the popular notion of "Man the Toolmaker," which regarded the fabrication of tools as the fundamental criterion of humanness. But, ironically, he was one of the very few paleoanthropologists who continued to reject an australopith ancestry for humans. A former devotee of the Piltdown "Eoanthropus," albeit rapidly turned skeptic, he preferred to believe that the genus *Homo* had extremely deep roots in time and would ultimately be found to have lived alongside the australopiths. Still, after toying briefly with the idea that the Olduvai australopith had been killed and eaten by a "more advanced form of man" that had made the tools found nearby, Leakey opted to conclude that his new hominid, which he assigned to the new genus and species *Zinjanthropus boisei*, had indeed been the toolmaker.

Yet the fossil about which Louis felt so conflicted was to be the Leakeys' ticket to fame and (by previous standards) fortune. In contrast to the staid professors who had previously pontificated about fossils found by others, the opinionated, exotic, and weather-beaten "white African" couple became the first hominid fossil finders to achieve international celebrity status—something that had eluded even the pathbreaking Eugene Dubois. This new public profile came about not least because the mighty publicity machine of the National Geographic Society, which now began to pour research funds into Olduvai, needed a good story. And it didn't have to wait long. In short order, a couple of leg bones were found in Bed I at a spot not far away from the *Zinjanthropus* site; then another locality, slightly lower in the section, yielded some hominid hand bones, most of the left foot, some bits of braincase, and finally a nice partial lower jaw with beautifully preserved teeth. Remarkably, the teeth in the mandible did not match up in the least with the upper teeth preserved in the *Zinjanthropus* cranium. The lower incisor teeth were much larger than you'd expect those of *Zinjanthropus* to be; the canine was much fatter; and the premolars and molars were much smaller (see figure on next page). Without question, this was another kind of hominid entirely.

At first Leakey was uncharacteristically reserved in his assessment of his new finds, but soon he was declaring that he had found something distinct from both *Zinjanthropus* and any South African australopiths. This was actually a bit of a stretch, since the teeth in the new mandible reminded many of the gracile australopiths from sites such as Sterkfontein, and the only substantive morphological difference Leakey found himself able to point to was a rounder outline of the premolar chewing surfaces. Still, such niceties didn't prevent him from asserting that not only was the new Olduvai hominid not an australopith, but that it represented "a new and truly primitive ancestor of *Homo*." This naturally enough didn't please South African colleagues who saw their graciles being relegated to a side branch of human evolution, but it did have the advantage of allowing Leakey to transfer authorship of the stone tools found in the lower layers at Olduvai from the embarrassing *Zinjanthropus* to the new hominid. At last, he had a candidate for the Earliest Toolmaker he could be happy with.

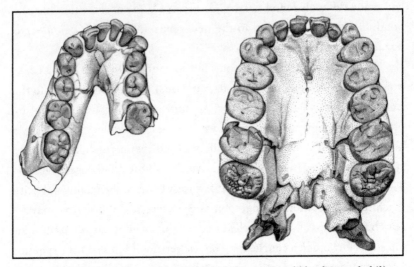

Two hominid dentitions from Olduvai Gorge. Left: type mandible of Homo habilis, *OH 7. Right: Maxilla of OH 5, type specimen of* Paranthropus boisei. *Although one of these dentitions is lower and the other is from the upper jaw, the contrast in size and proportions of the teeth is very evident. To scale. Drawn by Don McGranaghan.*

The tools in question were very simple ones. In Bed I, the stone tools eroding out were no more than fist-sized cobbles of quartzite or fine-grained volcanic rock, modified by a few blows with a stone "hammer." Sometimes the hammering was done while the core was loosely held in the hand, at others while resting on another rock that served as an anvil. Mary Leakey made artifacts of this kind her particular subject of study and eventually identified several different variants, depending on overall form and the amount of modification involved. Nowadays most prefer to view the sharp flakes struck off the core with each blow as the primary product of the tool-maker, and indeed Louis had become expert at rapidly skinning and butchering antelopes with flakes of this kind. In this interpretation, the shaped cores themselves are mainly by-products of obtaining flakes, although there is no doubt that many cores were also used for heavy-duty pounding tasks.

Back in the 1930s Leakey had dubbed this simple stone-working industry the "Oldowan," for the site at which it had first been identified; according to Mary, the basal industry gave way higher in the Olduvai section to the "Developed Oldowan," based on the proportion of cores of different shapes and the appearance of bifaces: stone cores flaked on two intersecting sides to produce an apparent cutting edge. Later studies have tended to question this difference, but evidence of true technological change was evident in the form of the handaxes that began to appear a little higher up in Bed II: implements much like those described from the Somme Valley by Jacques Boucher de Perthes. Associated with deliberately shaped tools of this latter kind was a relatively voluminous (1,067 ml) braincase, found in 1960 and known by the unromantic name of OH 9 (Olduvai Hominid 9). Louis was strangely reticent about this specimen, but most early commentators compared it to *Homo erectus*—even though it doesn't look an awful lot like the hominid from Trinil.

The mammal fossils found in Bed I along with *Zinjanthropus* and the other hominids appeared to be of early Pleistocene age. That meant they were pretty old, but what exactly that meant in terms of elapsed time was unknown. Louis himself had guessed around 600,000 years; although this estimate was more or less a shot in the dark, most of his colleagues felt it reasonable enough. So imagine the astonishment in 1961, when Leakey and the Berkeley geologists Jack Evernden and Garniss Curtis announced they had obtained an absolute date for the fossils of 1.75 million years! They had achieved this by applying a new method of dating, known as potassium-argon (K/Ar), to volcanic ash that interleaved the fossil-bearing sediments at the bottom of Olduvai Gorge.

Like radiocarbon, K/Ar depends on the decay of unstable radioactive atoms, although in this case it is the accumulation of stable daughter product that is measured in a rock sample, rather than the quantity of unstable parent isotope that remains. Radioactive potassium decays to the stable rare gas argon very slowly, with a half-life of about 1.3 billion years, so it is ideal for use on very old rocks. But in modern variants K/Ar has been successfully used on rocks as young as 100,000 years or even less, and volcanic rocks such as the tuffs (volcanic ashfalls) and lavas that intercalate the Olduvai sediments are ideal for dating by this method. This is because not all of the argon gas you find trapped in a mineral is necessarily formed by the decay of radioactive potassium—unless the rock has been heated to drive all argon out. The incandescent temperatures of volcanoes do just that and reset the radioactive clock to zero, so that the time when air-fall tuffs or lava flows were expelled from the parent volcano can be accurately dated. Fossil-bearing sediments just above or below a dated layer will usually be just a little younger or older than it is, and if those sediments are bracketed by two datable layers, maximum and minimum dates can be specified. K/Ar dating had originally been introduced—again, back in that fateful year of 1950—as a means of dating ancient salt deposits. But it had never previously been used in the time-range of early hominids; its deployment at Olduvai came as a revelation!

I am just a couple of years too young to have experienced at first hand the shock that the absolute dating of the lowest levels of Olduvai delivered to paleoanthropology. But when I first became acquainted with the field a few years later, in 1964, it was still in the process of coming to terms with the almost unimaginable ancientness of the earliest human fossils. Predictably, the immediate reaction of some experts was to believe that the new date (actually, the average of several separately made dates) made the Pleistocene far too long. Among the initial doubters was Ralph von Koenigswald, who recruited to his cause no less an authority than the German geochemist who a decade earlier had made the very first K/Ar date. Nonetheless, general acceptance of the Olduvai date followed quickly, and the K/Ar method and techniques descended from it have become indispensable parts of the paleoanthropologist's arsenal.

But Olduvai had plenty more surprises in store. It was rapidly confirmed that the foot bones were those of a biped, although the hand bones (not all of which, it turned out, were actually hominid) gave analysts plenty to argue about. Further, during the 1963 field season a site low in the middle part of Bed II yielded a partial cranial vault and associated upper and lower jaws, and a slightly older locality produced a very fragmentary skull with most of its teeth. The Leakeys considered that all of these bits belonged to the same kind of hominid as the "pre-*Zinjanthropus*" fossils they had earlier discovered right at the bottom of Bed I. What they hadn't found at Olduvai was a lower jaw to go with the *Zinjanthropus* specimen itself.

In 1964 a candidate mandible showed up some 50 miles away, in slightly later deposits (now known to date to about 1.4 million years ago) at Lake Natron. The height of the vertical part of the jaw indicated a much shallower face than the Olduvai *Zinjanthropus* possessed. But the characteristic dentition was there: tiny incisors and canines, followed by massive chewing teeth. By this time, incidentally, the Leakeys had demoted their *Zinjanthropus* to the status of subgenus (a fairly meaningless category) of *Australopithecus*.

So by the end of the 1964 field season it had already become clear that, just as in the South African cave fills, at least two different kinds of hominid were present in Bed I and the lower levels of Bed II at Olduvai (together dating, as we now know, between about 1.6 and 1.8 million years ago), while higher in Bed II (at around 1.2 million years ago) there was yet another hominid that was comparable in brain size to *Homo erectus* from eastern Asia. Up to that point, Leakey had been circumspect as to what he thought his pre-*Zinjanthropus* was, but in the spring of that year, in collaboration with the English anatomist and primatologist John Napier and the South African paleoanthropologist Phillip Tobias, he proposed that the more lightly built hominid fossils from Olduvai all belonged to an amazingly early species of the genus *Homo*. Because it was this hominid with which they associated the early stone tools of Olduvai, they dubbed it *H. habilis:* "handy man." Almost certainly, it was this association with stone tools that gave Leakey, Tobias, and Napier the courage to place the new materials in *Homo*, in obeisance to the "Man the Toolmaker" preconception.

As we saw, the teeth in the original lower jaw looked much more like those of gracile South African *Australopithecus* than anything else then placed in our own genus. And other morphological features didn't help much to argue the case for inclusion in *Homo*. Save for one thing. The year before, Tobias had reviewed brain sizes in South African *Australopithecus* specimens and had come up with an average of a little over 500 ml. Once pieced together, the lowest Bed I Olduvai cranial fragments suggested to Leakey a slightly (and to him significantly) bigger brain. His first estimate was 680 ml, while reconstructions of the other fragmentary skulls have subsequently come in a bit lower but nonetheless well over the 500 ml mark. Still, a marginally larger brain, under half the size of ours today, was hardly a very robust justification for the huge extension of the genus *Homo*, in both time and morphology, that the creation of *Homo habilis* implied. Certainly, one would be hard put to point to any other mammal genus encompassing a similar range of brain sizes—although rapid brain size increase is in itself a vanishingly rare phenomenon.

It actually took the discovery of fossils elsewhere in East Africa to convince a reluctant paleoanthropological establishment, rightly or wrongly, that the new species *Homo habilis* was a reality to be contended with. But when it came, that acceptance had profound implications for the future. This was because the taxonomic distinction unconsciously placed the East and South African hominid fossil records in separate compartments of the paleoanthropological mind. Gracile *Australopithecus* was a South African phenomenon, and early *Homo* was an East African one. And while this absolute separation did not last very long, it planted the seed for a mind-set that lingers in attenuated form even today.

Meanwhile, Mary Leakey's continuing archaeological work at Olduvai was giving *Homo habilis* an increasingly human face. A "rough circle of loosely piled stones" at a site low in Bed I was interpreted as a possible windbreak—the world's earliest deliberately constructed shelter. For all his primitive features, Handy Man had possessed a home! Later work suggested that the ring actually consisted of stones shattered and moved by the expanding roots of a tree, but at the time the additional cultural complexity implied by the windbreak was seen to support the notion that early *Homo* had roamed the East African savannas a very long time ago. Even more influentially, in June 1964 Tobias and Koenigswald met in Cambridge to compare the new Olduvai material with the still poorly dated collection of *H. erectus* fossils discovered by the latter in Java before the war. This was one of the very few occasions on which original hominid fossils from two continents have ever been seen directly side by side.

The two scientists came from different generations and backgrounds, but somehow they contrived to converge on a rather woolly statement, just as McCown and Keith had done at Mount Carmel. Basically, Tobias and Koenigswald proposed that a single unitary sequence of human evolution had been manifested in both Africa and Asia. The first "grade" of this sequence was epitomized by gracile *Australopithecus* in South and East Africa, and it was placed in Asia with a question mark. Even then, it remained equivocal just what the authors meant by *Australopithecus,* and especially by its East African expression, and Tobias and Koenigswald went on to express doubt that there was "as yet any unequivocal evidence pointing to the presence in Asia of a frankly australopithecine grade of organization." More confidently intercontinental, Grade 2 consisted of Olduvai Bed I *Homo habilis* in East Africa, the *Telanthropus* stuff from Swartkrans in South Africa, and maybe also of a couple of jaws from Java that Weidenreich had earlier dubbed *Meganthropus palaeojavanicus.* Grade 3 encompassed some, at least, of the Bed II *H. habilis* material, plus the balance of Koenigswald's *Pithecanthropus* material from Java. Grade 4, which the authors explicitly declined to discuss in any detail, consisted of OH 9 in

East Africa and Dubois' original Trinil hominid and the Peking Man materials in Asia. It also included *Atlanthropus*, a genus created in 1950 by the French paleontologist Camille Arambourg to contain three mandibles from Tighenif, in Algeria.

Here we see good evidence for persisting conflict in the international arena between older ways of thinking and newer ones. Mayr's impact on English-speaking paleoanthropology had been encompassing and immediate. But it took a while for his views to catch on elsewhere, and it wasn't until the early 1970s that the Synthesis really took hold in most countries of continental Europe. The overall vision hammered out between the younger, English-speaking Phillip Tobias and the more traditional, German-speaking Koenigswald basically reflected the new view of evolutionary process. But the wild profusion of names used in their review, and the uncertainty as to exactly which fossils those names applied, clearly exemplified Koenigswald's prewar preference for giving everything in sight a new name, with splendid disregard for any systematic implications. Given this, I'm pretty sure both scientists must have been unhappy with the published fruit of their joint labors—though I never heard either of them say this explicitly. But I am equally convinced that their sibylline utterances reflected a shared perplexity in the face of the fossils they confronted. Tobias and Koenigswald were truly breaking new ground in making their intercontinental comparisons. What is more, there was not yet a huge amount of context within which to arrive at any conclusions at all.

CHAPTER 5

EVOLUTIONARY
REFINEMENTS

I ARRIVED IN CAMBRIDGE TO BEGIN MY UNDERGRADUATE STUD-
ies a couple of months after Tobias and Koenigswald had left, in total igno-
rance of what had just transpired in the very laboratory I now haunted. My
interests in anthropology had been piqued by my childhood experience of
East Africa—I had actually been in boarding school in Nairobi, just down
the road from where they lived, at the very moment Louis and Mary Leakey
were making their epic discovery of *Zinjanthropus*. But although I'd met
many cultural anthropologists at Makerere College in Uganda, where my
father worked, I remained blissfully unaware of the biological end of the
science until after I arrived in Cambridge. Still, once exposed to biological anthropology I was hooked, especially after David Pilbeam joined the
faculty at the beginning of my second year. Pilbeam was a Cambridge whiz
kid who, after graduating, had gone off to do a doctorate at Yale with Elwyn
Simons, a rising star of primate paleontology. Such is the inbred nature of
academic genealogies that Simons had himself studied at Oxford with Wilfrid Le Gros Clark, rehabilitator of the australopiths. Pilbeam was a simultaneously intimidating and inspiring presence who exemplified the New
Wave in paleoanthropology, and his impact on both me and his profession
was huge.

At the time, both Pilbeam and Simons were—as now—more interested in fossil apes than in fossil humans; but in 1965 they collaborated on an analysis of the early human fossil record that perfectly embodied the new perspective on evolutionary process, and contrasted dramatically with the Leakeys' authoritarian approach. "Perhaps," they wrote, "we have been looking down the wrong end of the telescope [by] . . . looking backward . . . from the vantage point of the Recent." So "one must not ask 'How *Homo*-like is *Australopithecus?*' but rather the opposite." In a long and detailed discussion of the earlier phases of the hominid fossil record, Pilbeam and Simons covered all the bases that Mayr had touched on; in particular, they followed his insistence that issues of ecology and behavior had to be added into the interpretive mix from the very beginning. As I realized much later, this order of business was actually a mistake, because you really do have to properly appreciate just who the actors—the species—are before you will ever understand the evolutionary play you are trying to unravel. But in historical terms this was a truly pathbreaking contribution that defined the priorities of a burgeoning new generation of paleoanthropologists. For the first time—much to Leakey's chagrin—Pilbeam and Simons were addressing the issue of *Homo habilis* and the South African australopiths in a global way, and in the light of the New Evolutionary Synthesis.

Dating of the South African materials was at that point still quite shaky, as was that of the later strata at Olduvai. These uncertainties led Pilbeam and Simons to believe that the *Homo habilis* fossils from Beds I and II at Olduvai Gorge might have been separated in time by as much as a million years. This perceived (and, as it later turned out, excessively large) time gap enabled them to make a significant distinction between Bed I and Bed II *H. habilis*. And that, in turn, allowed them to associate the Bed I stuff with South African *Australopithecus africanus,* while comparing the Bed II materials to John Robinson's *Telanthropus* fossils from Swartkrans. Nonetheless, in good Mayrian fashion, they also perceived strong similarities between the earlier and later materials, enabling them to envisage a single continuum of hominid evolution that had been represented in both East and South Africa. If *H. habilis* was properly positioned within the genus *Homo,* they said, then a southern and eastern African form that by systematic convention had to be called *H. africanus* had gradually given way to *H. erectus,* and finally to *H. sapiens* via intermediates they didn't identify. Pilbeam and Simons conceded that, although the robust australopiths had small incisors and canines—and thus presumably prepared their food using tools—they nonetheless constituted a separate and parallel lineage. Nonetheless, they had basically contrived to incorporate all of the new fossils found over the last decade into Mayr's original scheme. No longer were the gracile australopiths and early *Homo* the separate and parallel lineages that Leakey wanted them to be.

This affirmation of a rigidly linear scheme of early human evolution perfectly complemented a spectacularly assertive attack that had been made the previous year on all who would exclude the Neanderthals from modern human ancestry. According to its author, the paleoanthropologist C. Loring Brace, human evolution had consisted of a stately, and by implication almost inevitable, progression from australopiths through "pithecanthropines" and Neanderthals to modern humans. Anyone who thought otherwise was branded "anti-evolutionary." Brace based his argument on his perception of two dominant trends in human evolution. The first of these was steady enlargement over time in the size of the brain—which, if not too closely examined, certainly seemed to fit a continuum. Two million years ago, hominid brains were barely, if any, bigger than those of chimpanzees today. A million years later they had on average doubled in size, and today they have increased by almost as much again. The second trend Brace noted was the progressive reduction of hominid faces and the teeth they contained, something he attributed to a steady refinement of cutting tools that relieved the dentition of many of its former duties. Although many complained about Brace's brash style, there was a lot of sympathy for his argument, and in laboratories all over America the dust was brushed off Aleš Hrdlička's 1927 broadside to the Royal Anthropological Institute. The "Single Species Hypothesis" was back in business.

But Pilbeam and Simons had an agenda that went well beyond simply arranging hominid fossils in a linear sequence. They wanted to create a comprehensive account of human evolution that encapsulated all of the elements Mayr had thought important. Their chosen story started with the reduction of the hominid canine tooth. By 1964 it was already clear that this reduction had largely been accomplished (as Darwin himself had predicted) by a very early point in the human story; and the small canine was not only a remarkable feature that united all the hominids, but it also had major consequences for numerous aspects of their behavior. Apes retain long, slashing upper canine teeth that are stowed between the back of the lower canine and the front of the first premolar when the mouth is closed. Ape upper canines even sharpen their rear edges against the lower teeth behind them. When the mouth is open and the sharp canines are exposed, the ape thus has a very potent weapon at its disposal if threatened by a conspecific or attacked by a predator (chimpanzees are even rumored to have killed leopards). Pilbeam and Simons also observed that apes can use their large canines for such heavy dietary duties as stripping bark and otherwise "nipping, tearing and shredding vegetation." That is the upside of these big teeth. The downside is that, when the mouth is closed or nearly closed, the lower and upper jaws are basically locked together by those projecting canines. The lower jaw cannot move from side to side, as you might want it to do when grinding the tough foods that are thought to have constituted at least part of early hominid diets. Their large canines more or less limit apes to a chomping action.

In line with their emphasis on behavior and ecology, Pilbeam and Simons tackled this functional issue early in their argument. They observed that any primate with reduced canines and incisors would have been limited in its ability to process vegetation. And they suggested that this deficiency had to be compensated for by the *ad hoc* use of cutting tools. So by their reckoning, any hominid with a reduced canine *had* to have used tools. And if it did, it obviously already possessed what was perhaps the most fundamental quality of humanity. Even Louis Leakey would have had to agree with that. What's more, once hominids had become committed to the ground, the freeing of the forelimbs from the demands of locomotion would have enhanced this tool-using proclivity.

But although it certainly enhanced the australopiths' claim to humanity, this long chain of reasoning was not produced primarily for that purpose. Instead, it was principally aimed at reinforcing the claim to hominid status of a much earlier primate: *Ramapithecus brevirostris*. Mainly known from a fossil half palate discovered in the Siwalik Hills of India (Hugh Falconer's old stamping ground) before World War II, *Ramapithecus* as reconstructed by Simons in 1961 was remarkably humanlike. It had small incisors and canine teeth; its upper tooth row was set in an arc, as in modern people; it had broad, simple chewing teeth; and its face was short from back to front. This apparently flat muzzle was a feature of special interest, because in Pilbeam and Simons' view it was strongly correlated with upright posture, a more or less globular head being easier to balance atop a vertical spine. Uprightness was, of course, associated with bipedality, and by extension with hands that were free to wield tools. So since "assignment to Hominidae can reasonably be made for all those species that show evolutionary trends toward modern *Homo,* whenever those trends appear," *Ramapithecus* was inevitably a hominid.

The influential paleontologist George Gaylord Simpson had already expressed the view that the origins of distinctive "higher taxa" (groups larger than the genus), such as Hominidae, could generally be traced to major changes in "adaptive zone"; and the complex construct described by Pilbeam and Simons certainly signaled that a major adaptive shift had announced the advent of Hominidae. What was most remarkable, though, was that *Ramapithecus* was then thought to be 12 million years old! The roots of the hominid family evidently ran very deep indeed, back into the Miocene epoch.

All this added up to a compelling story, and it certainly impressed me when I duly went to Yale to study under Simons in 1967, a year before Pilbeam returned to join the Yale faculty. Not long after arriving I took a course with the great ecologist G. Evelyn Hutchinson, during which I examined the fossil fauna that had been found alongside *Ramapithecus.* What I found was that the animals concerned had been broadly typical of forested settings, suggesting that *Ramapithecus* had lived a principally arboreal life. That's as far as my term paper on the subject went; but so strongly did everyone around me believe in the intimate relationship between morphology and ecology and behavior that the version eventually published reads like a paean to the evolution of upright posture and tool use in the trees.

Louis Leakey was of course appalled, although less by the more extreme claims made in the published article than by the pretty obvious idea (which he himself had actually exploited when it had suited him) that the mammals associated with a fossil of interest might carry a signal about the environment it had lived in. But he also had his own agenda, in the form of an alternative candidate for earliest hominid. This was *Kenyapithecus wickeri,* a species based on a piece of upper jaw from a site in Kenya that was dated to about 14 million years ago. And what particularly got his goat was that Pilbeam and Simons had shown the temerity to assert that his cherished *Kenyapithecus* was the same thing as their *Ramapithecus.* Even worse, if the upstarts were right, the Indian fossil's name had priority. Still, whether or not the two were exactly the same thing, for many observers the perceived similarities between the Indian and Kenyan fossils cemented the notion that the last common ancestor of apes and humans had lived more than 14 million years ago. The family Hominidae evidently had a very long history.

What is more—especially after some fragmentary fossils from Europe and China had been added to the mix—those same similarities also hinted that human evolution had been a worldwide and gradual phenomenon of the sort that had recently been described by Tobias and Koenigswald, and long before them by Weidenreich. As a result, many became persuaded by the mid-1960s that, in the words of Pilbeam and Simons, "the commitment to a hominid way of life had been made by the late Miocene, and our earliest known probable ancestors . . . might already have adopted a way of life distinct from that of their ape contemporaries."

This, then, was the view of human evolution I was initially taught, in my impressionable early and middle twenties. The human family went back a really long way, and it was doing some very distinctive things, including making tools (never actually found, by the way), a very long time ago. And like everyone else I was also thoroughly seduced by the elegant simplicities of the Synthesis, and by the progressive view of human evolution that flowed from it. Human beings love good stories, and at the level of narrative the Synthesis told a very good tale indeed. What could be more the stuff of drama than the story of a plucky population single-mindedly battling its way, against all ecological odds, from a lowly primitive status to its present exalted station? After all, as the science historian Misia Landau pointed out, this narrative of human evolution shares its basic plot elements with some of the world's most durable folktales. What is more, at least in principle, this scenario also offers the elements of a comprehensive scientific explanation, for with a little bit of tunnel vision it is entirely possible to visualize virtually any evolutionary phenomenon you might care to mention as the product of gradual gene frequency changes under the guiding hand of natural selection.

Unfortunately, however reductively satisfying the Synthesis might be, it says nothing prescriptive about how one should actually approach the practical task of figuring out how individual evolutionary histories had unfolded. So, while it provided a framework within which lineages could be seen to evolve, it did not provide an aspiring paleontologist with much of a procedural guide. And as a result, although in theory graduate school ought to be a sort of apprenticeship during which one learns the tricks of one's anticipated trade, in practice I ran into difficulties understanding exactly what those tricks were—even though I was in a privileged position from which to observe. Elwyn Simons was my adviser, and a brilliant finder of fossils. He was a genial mentor who knew how to stand back and allow his students to follow their noses. And he was a highly productive scientist who, chaotically but without apparent effort, published paper after paper in the most prestigious journals, describing and explaining the significance of his fossil discoveries. But just how he pulled off this last trick was far from obvious to a naïve eye like mine. Indeed, I suspect that even if I had dared ask him outright exactly how he reached the conclusions he published, he would have had difficulty explaining it to me. It was just what he did—and did very well.

With the benefits of hindsight, it is not hard to see why the recipe for analysis in paleoanthropology was so difficult to pass along. The fact was, there really wasn't one. By the time the 1960s rolled around, paleoanthropologists certainly had a consistent concept of evolution within which to situate the fossil forms they studied, and the mechanisms by which the evolutionary changes they saw seemed soundly established. But when it came to analyzing the actual tangible fossils themselves, and particularly to discerning their identities and relationships, everything still depended on intuition and expert judgment. And just how to develop intuition is a hard thing to teach.

Having later had plenty of opportunity to observe just how powerfully the mind-sets inculcated by strong-minded professors can permanently condition the way their students see the world, I am enduringly grateful for Simons' hands-off approach. Still, at the time it didn't seem to help much. I would regularly sit in my "office" in a basement storeroom at Yale's Peabody Museum of Natural History, watching in bemusement as a succession of visiting researchers spent days poring over the fossils they retrieved from the storage cabinets. All the time, I was wondering what on Earth they were actually doing as they scribbled page after page of notes. How were they extracting the messages about the history of life that were encoded in those fossils? For a long time I lacked the effrontery or the courage, whichever it might have been, to ask. When finally I did get it together to inquire, I was massively disappointed. "You look at the fossils long enough," I was told by a distinguished professor, "and they speak to you." Having never been personally addressed by any fossil, no matter how long I had stared at it, I never asked again. The problem was, of course, that no matter how sophisticated evolutionary theory had become, the era of authoritative pronouncement was still upon us. The only real difference from the old days was that, certainly in the case of human fossils, such pronouncement was by then generally in service of the seductive linearities of the Synthesis.

Perhaps the best thing that could have happened to me, then, was my entirely accidental selection of the then rather neglected "subfossil" (recently extinct) lemurs of Madagascar as a thesis topic. As I mentioned right at the start of this book, the lemurs—a unique group of relatively primitive primates that diversified in isolation on their great island redoubt—are astonishingly diverse. Without competition from monkeys or other primate groups, and with a multitude of ecological opportunities, the lemurs radiated over the course of 50 million years or so to do virtually everything you could imagine a primate doing. They employ every conceivable way of locomoting around in the trees; they exploit virtually every resource that a forest environment can offer; and they span the gamut of primate social organizations. The result is a riotous profusion of species, grouped into five families and ranging in size from the two-ounce mouse lemurs to the gorilla-sized and recently departed *Archaeoindris*. So obvious is lemur diversity, indeed, that as I've noted it is almost the first thing even the most casual observer notices about the group.

But even more significantly, although the variety of lemurs in Madagascar is particularly remarkable, diversity itself is hardly an unusual phenomenon. As I looked around for other mammal radiations to compare with the one the lemurs had produced in Madagascar, I rapidly came to see that diversity is actually the norm for any widespread group, certainly one as successful as Hominidae. The linear model of human evolution favored by my advisers—and that, remember, Mayr had felt it necessary to justify by laborious special pleading—was actually a striking exception to the general rule. To members of a single species that dominates the world today, it may seem natural to reconstruct our biological history by simple extrapolation back in time; but, as we will see, this perspective of ours is only made possible by some highly unusual circumstances.

MOLECULES

It took a while for me to absorb the paleoanthropological implications of these realizations; and anyway, by this point I was becoming increasingly preoccupied by my work on the extinct lemurs. Meanwhile, part of the human evolutionary edifice I had been taught was already running into trouble, largely due to the intrusion into paleoanthropology of "molecular systematics." In the early 1960s the molecular geneticist Morris Goodman, of Wayne State University, began to compare primate species using molecular criteria rather than the gross anatomical features that were the stock-in-trade of human paleontologists. Initially, Goodman resuscitated an approach that had been employed by the English bacteriologist George Nuttall around the turn of the twentieth century. Nuttall had compared the strength of the cross-reactions among pairs of species to each other's blood proteins, the idea being that closely related species would show stronger immunological reactions. He eventually tested a wide variety of animals using this approach, and found that humans and apes indeed had a stronger "blood relationship" than either had with Old World monkeys.

Though interesting, that was hardly a surprise, and there the matter rested until Goodman came along half a century later with more refined techniques at his disposal. This time, the results were more unsettling to received wisdom. Conventional classification placed humans in one family, Hominidae, while all of the great apes were grouped together in another, called Pongidae (for the orangutan, *Pongo*, the first ape to receive a zoological name). Yet Goodman found that humans and the African apes— gorillas and chimpanzees—were much more similar to each other than any of them was to the Asian orangutan. Indeed, so similar did Goodman find the African apes and humans that he placed all of them together in the family Hominidae, leaving the orangutan to occupy Pongidae all by itself.

The idea that we humans are more closely related to some apes than to others was pretty strong stuff. After all, if all the others were apes, and we fell somewhere in the middle, then we humans were by definition apes as well. But Goodman didn't leave it at that. He followed up his initial inquiries with a series of studies of various blood proteins using electrophoresis, a technique that sorts molecules according to their size and weight. Again, he found the strongest similarities between humans and African apes, although in some molecules the chimpanzees came out closer to humans, while in others gorillas did. Predictably enough, the initial reaction of traditional primate systematists was to flatly reject his findings, and Goodman himself was worried by the incompatibility of his molecular results with the traditional morphological classification. Well, as a morphologist I'm embarrassed to admit it, but it eventually turned out that the inconsistencies were due more to deficiencies in the ways morphologists treated their data than to any problems with the molecules.

Less concerned by the failure of the molecules to track received wisdom were two Berkeley biochemists, Vincent Sarich and Allan Wilson. In 1966 and 1967 these pioneers published studies that employed a quantitative technique to compare the blood proteins known as albumens across a range of primate species. The numbers they obtained enabled them to construct a family tree among the species they examined—a tree that basically agreed with Goodman's. But Sarich and Wilson went further. According to them, their data revealed that the albumen molecule changed at a constant rate. By using just one generally agreed calibrating date from the fossil record (30 million years for the divergence between humans and Old World monkeys), they were able to calculate on this assumption that humans and chimpanzees had last shared a common ancestor a mere 5 million years ago! Sarich and Wilson later softened a bit on the date. But not by much, and in 1971 Sarich, with scant regard for the tender feelings of the paleontologists, wrote that "one no longer has the option of considering a fossil older than about eight million years as a hominid *no matter what it looks like*" (emphasis his).

None of this looked good for *Ramapithecus*—about which, as it happened, some paleontologists were also beginning to harbor doubts. As more material was discovered in Africa and Asia those doubts grew, and one by one the features that supposedly linked *Ramapithecus* with hominids were discarded, until virtually the only humanlike trait shared uniquely by *Ramapithecus, Australopithecus,* and *Homo* was a thick coating of enamel on the molar teeth. Then, even this was found to be present in an Indian Miocene orangutan relative known as *Sivapithecus*. The coup de grâce for the wannabe hominid eventually came in 1980, when the English paleontologist Peter Andrews and his Turkish colleague Ibrahim Tekkaya formally subsumed *Ramapithecus* into the genus *Sivapithecus,* which a new fossil from Turkey showed had possessed a face very reminiscent of that of an orangutan. If the orangutan-like *Sivapithecus* was an ape, they said, then so was *Ramapithecus.* What's more, by this time even Pilbeam was beginning to believe that *Ramapithecus* was an ape. In 1976, his own expedition to Pakistan had recovered a lower jaw of *Ramapithecus* that had a suspiciously apish curve to it; and in the 1979–80 season the expedition recovered a splendidly preserved face of *Sivapithecus* that looked even more orangutan-like than its counterpart from Turkey. This latter discovery tipped the balance for Pilbeam. Bravely, he did an abrupt about-face on the *Ramapithecus*-as-hominid issue, in the process earning himself widespread admiration for his intellectual honesty and open-mindedness. This was a remarkable feat in a profession whose practitioners are often slow to change their minds, even in the face of compelling evidence.

Inevitably, the finding that humans nested somewhere within the African great ape group raised the issue of exactly what it meant to be a member of the family Hominidae. If you went with Goodman, you could simply shift the gorillas and chimpanzees to Hominidae—or even, if you agreed with some zealots, to the genus *Homo*. But if you did that, you would be disguising the very important fact that members of the human group were doing something radically different from any of the apes, which all remained much more conservative in their lifestyles— wherever they lived, in Africa or in Asia. In traditional taxonomy there are no clear rules to govern cases of this kind. Ideally, you want every group you recognize to be monophyletic, which is to say, descended from the same common ancestor; but there is no hard-and-fast procedure for de- termining at what level in the taxonomic hierarchy such groups should be recognized. One option in this case might be to create three subfamilies of Hominidae—Homininae for australopiths and *Homo*, Ponginae for the orangutans, and Paninae for the chimpanzees and gorillas. But the three subfamilies would not be equivalent, meaning that two of them are inevi- tably more closely related to each other than either is to the third, despite their equality of rank in the taxonomic hierarchy. You could subdivide again, but things then start to get a bit clumsy; so most paleoanthropolo- gists today are happy to go with the subfamily Homininae for australo- piths and humans, while leaving it to the ape experts to worry about how to classify their primates of interest.

From my point of view, however, it seems equally legitimate to adopt a "bottom-up" perspective, and to take account of the fact that what we have traditionally called hominids are in fact amazingly diverse: indeed, even quite conservative paleoanthropologists would nowadays recognize a couple of dozen extinct human species (see the family tree on page 217). Based on this variety, the group that contains humans, plus all their extinct relatives that lived after the common ancestor, amply deserves recognition at the family level. So, arbitrarily—but not more arbitrarily than any of the other options available—I am going to continue to use the term "hominid" for this group.

While we're on terminology, it might be appropriate to ask where this leaves that tricky term "human." As we saw earlier, this word has been freely bandied about since well before anybody had any idea that creatures formerly existed that were closer to us than the apes are, so it has typically been used pretty loosely. Even the same person will use it differently in different contexts. In my case, by the term "human evolution" I refer to the evolution of the entire hominid family. But I would describe a hominid as "human" only if it belonged to the genus *Homo*. And the sole "fully human" species is our own, *Homo sapiens*.

MODELS AND MONKEYS

One of the most influential and innovative papers in paleoanthropology published during my time in graduate school was written by New York University's Clifford Jolly. Back in the 1950s, John Robinson had surmised that the morphological differences between the gracile and the robust australopiths had been driven by dietary specialization: that the graciles had been omnivores, with some meat in their diet, while the robusts remained vegetarians, perhaps with a preference for tougher foods such as plant tubers. Now, Jolly tested speculation of this kind by looking for a "living model" that might give some perspective not simply on this matter, but also on the origins of bipedality among hominids, and what might have driven the differences between hominids and apes. A specialist in the behavior and genetics of baboons, Jolly had been working in a region of Ethiopia where more generalist hamadryas baboons hybridized, in a limited area, with very specialized open-grassland gelada baboons. And he saw significant behavioral as well as morphological differences between the two species that he thought he could tie in with their dietary preferences.

Looking at the anatomical structures of the two, Jolly also noticed that some of the ways in which geladas differed from hamadryas paralleled those in which hominids differed from apes. And he associated these differences with contrasting environmental preferences: of apes and baboons for more closed habitats, and of geladas and hominids for more open ones. Jolly pointed out that both geladas and humans have relatively short fingers, enabling more precise manipulation of objects at the expense of strong

gripping abilities. In the case of geladas, the objects most commonly ma-
nipulated were the grass blades, seeds, and rhizomes that provided the bulk
of their diet out there in the grasslands. Unlike hamadryas, which feed "tri-
pedally" in the open, standing on three legs while foraging with one hand,
geladas obtain these delicacies while sitting on the ground, holding their
trunks erect and feeding with both hands. To move from one feeding spot
to another nearby, they will often "bottom-shuffle," keeping their trunks
erect instead of getting up and moving on four legs.

Another similarity between geladas and hominids that Jolly thought
merited particular attention was a relatively short face, associated with a forward placement of the muscles that move the jaw and a limited gape within
which to accommodate food items while chewing. These seemed to be linked
to a reduction of the incisor teeth and a modest expansion of the molars in
both hominids and geladas, relative to those of their forest-living relatives.
These features, and more, were in Jolly's view all part of a single "small-
object feeding" functional complex that was related to the procurement and
mastication of small, tough nutriments. And he noted that, if this was the
case, incisal reduction in the hominids was not necessarily related to the use
of tools to obtain and prepare food, as Pilbeam and Simons had suggested.

Based on his deductions from diet, Jolly proceeded to articulate a specific scenario of early hominid emergence. In a first phase, he proposed, a
population of primitive apes had left the forest, attracted by the resources—
all of them small, and most of them tough—that were available in grassland
environments, and for which there was little competition. The features they
would have evolved to cope with this diet then "pre-adapted" them for the
next phase of hominid evolution, which involved the gradual incorporation
into the diet of meat that would have been obtained by males, while females
and juveniles continued to forage. Thus would have begun an economic
as well as reproductive separation of roles between the sexes, with myriad
social and ultimately cognitive consequences. Not all hominids would have
shifted out of the first phase, however. The hominids that took the plunge
gave rise to the gracile australopiths, and ultimately to our genus *Homo*.
The robust australopiths, on the other hand, were the end result of refinement of Phase 1 adaptation.

This was an extremely ingenious argument, and although it is possible to see, with the benefit of almost half a century of hindsight, that it
was flawed in numerous details, it is impossible to overstate the historical

importance of Jolly's study in inaugurating a new style of paleoanthropology that paid close attention both to functional anatomy and to the morphological integration of the organism in the attempt to reconstruct past hominid life. As we will see, the living model approach, along with Jolly's concern with open-country monkeys and their habitat, has since proven extremely productive. And finally, in the context of the origin of bipedality, Jolly's stress on holding the trunk upright, rather than on locomotion per se, has proven highly prescient. It is becoming increasingly appreciated that we are bipeds not because a remote quadrupedal ancestor improbably stood up, but because a suspensory ape, which moved around in the trees with its trunk erect, simply felt most comfortable standing and moving upright when it came down to the ground.

A NEW VISION

After four years at Yale, I had the great fortune to find employment in 1971 as a junior curator at New York City's American Museum of Natural History. By the time I had graduated I already knew enough to realize that I had not completely learned how to figure out the relationships among the fossils I studied. I was becoming more confident in making judgments, but I still found it hard to put my finger on exactly what I was doing. And for a while, it actually didn't seem to matter that much. I had scraped by with a thesis on a family of extinct lemurs, and one advantage of choosing this arcane subject was that I'd had to travel to Madagascar to see the fossils in the collections of the Académie Malgache. While I was there, I leapt at the opportunity to wander around in the island's forests and observe the lemurs that still occupy them. Instantly, I fell head over heels for these charismatic creatures, and my future plans soon began to revolve around the living lemurs rather than around their extinct relatives. I wanted to know more about the lives of these remote relatives—very poorly understood at the time—and I wanted to understand the basis of their diversity. I wanted to discover exactly which kinds of lemur lived where on this environmentally diverse island, and how they did it, and why they were where they were. Following this new interest involved embarking on field research very different from the fossil-oriented world I'd inhabited so far.

For better or for worse, my career as a field lemurologist didn't last very long. In 1972 Madagascar's postcolonial government was overthrown, and by mid-1974 the island had descended into chaos. Before the end of that year I'd been escorted out, which was why I had first found my way to the Comoro Islands where this book started. And by the middle of 1975 I was back in New York, with little immediate option but to return to the hominid fossils I had abandoned just a few years before. Fortunately, the moment could not have been more propitious, for in the early 1970s a much quieter revolution, this time in systematics, was underway right there at the American Museum.

Back in 1950 a German entomologist called Willi Hennig had published a pathbreaking work on systematic procedure, in which he outlined a very explicit approach to biological classification and the reconstruction of evolutionary relationships. The omnipresent Ernst Mayr later called this approach cladistics (from the ancient Greek word for branch, since it used branching diagrams to represent relationships). But it was only after 1966, once Hennig's book *Phylogenetic Systematics* appeared in English, that it got much attention outside Germany. That it got noticed at all in the English-speaking world was due in large part to a pair of American Museum ichthyologists, Donn Rosen and Gareth Nelson, who indefatigably promoted the new viewpoint at a time when most systematists embraced one version or another of Wilfrid Le Gros Clark's "total morphological pattern" approach. Relationships among organisms were conventionally evaluated by intuitive assessment of overall similarity; and in the case of disagreement among authorities, intuitions were very difficult to compare. Choosing among plausible alternatives became a matter of whose pronouncement one instinctively regarded as more persuasive, and salesmanship was at a greater premium than rigorous reasoning was.

Although many traditional systematists would later claim that it was what they had been doing all along, Hennig's approach was in fact entirely novel, and it provided an explicit framework within which competing hypotheses of relationship could be compared, at least in theory. Hennig's most important insight was that not all morphological similarities are created equal—or at least, they don't stay that way. Systematists had, of course, long known that not all characteristics were of equal use in forming theories of relatedness, and they were always careful to distinguish "synapomorphies" inherited from a common ancestor—which do tell you something about relationships—from "convergences," which involved the independent (and thus systematically noninformative) acquisition of characteristics that looked alike to the eye. The wings of bats and birds are a classic example of the latter. But Hennig went further, and pointed out that even ancestrally inherited similarities are not all equal in systematics. Because once a feature has been used to determine common descent from a particular ancestor, it is no longer useful in establishing relationships among its descendants—after all, they all have it, if they haven't lost it. You thus have to look for common possession of other traits, or alternative versions of the same trait, to determine relationships within the group.

For example, possessing four legs tells the systematist that an animal belongs to the tetrapod group that all land-living vertebrates belong to. But it says nothing about which tetrapod groups are most closely related to each other. Consequently, you have to discard this unifying similarity when weighing relationships *within* the tetrapod group. To put the matter another way, similarities inherited from a common ancestor—whether they are morphological or molecular—may be "primitive," or they may be "derived," depending on how immediate that ancestry was. Any ancestral morphology (those four legs, for example) can be a derived character of the ancestor that unites a diverse group and distinguishes it from others. But within the group itself that feature will be primitive, and—unless it is lost—all descendants will have it. As a result, it is not useful for determining within-group relationships and should not be considered further. The bottom line is that evolutionary relationships can only be determined on the basis of shared derived character states; and if you view all similarities as carrying equal weight, you place yourself at risk of being misled.

So, how do you determine which character states are primitive for a group, and which are derived? You can check for a number of things. Most importantly, primitive character states will be distributed widely within your group of interest, while derived ones will be rarer. Even better, if you can find a particular character state in a close relative outside the group you are considering, that is a pretty good hint that it is primitive and cannot be used to link subgroups. Often, looking at how a character develops will also help: the fact that both fish and humans possess gill slits very early in embryonic life is an excellent tip-off that gills are primitive among vertebrates today (though they were derived in the ultimate common ancestor). More controversially, paleontologists sometimes assume that features appearing early in the history of a group are primitive, while those showing up later are derived—although this assumption is sometimes rather risky.

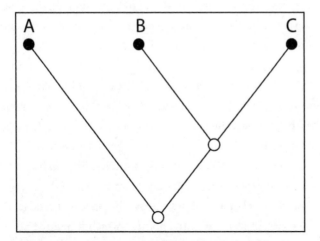

*An elementary cladogram. The solid circles represent three
actual taxa, for example three living species. The lines indicate
the genealogical relationships reconstructed among them.
The open circles at the nodes acknowledge that this is simply a
hypothesis of relationships, based on the shared possession of
character states by B and C that are not present in A. In other
words, the nodes do not correspond to any known taxon.*

To create your statement of relationships—otherwise known as a
cladogram—you take together all your species of interest, living or extinct.
Using those character state distributions, you then infer what the state of
each character was in the common ancestor of the group, and in the ancestors of all the subgroups within it. The actual ancestors remain hypothetical—you don't have to know them as fossils, for example—but the
character state(s) you infer for those ancestors are explicit. The result is an
unambiguous statement about the relatedness by ancestry of all the organisms in the group you are looking at. And your cladogram of relationships
within the group can potentially be falsified—rejected—or modified by the
addition of other morphological characters, or of extra taxa, to your analysis. What's more, the cladogram you come up with can be evaluated relative
to competing hypotheses of relationship by comparing the distributions of
the characters used to create each one.

As the cladistic approach began to bite in systematics, various procedural complexities became evident, perhaps the most significant of them

being that convergences—technically known as homoplasies—are much more widespread in Nature than anyone had imagined. Independent acquisition of characteristics that look similar to the eye is in fact rampant, introducing a huge amount of noise into the histories of relationship that systematists were trying to reconstruct. This was and remains a headache for anyone trying to produce a cladogram by eye; pretty soon, with personal computers beginning to appear on everyone's desk, complex algorithms were developed that allowed systematists to consider numerous characteristics simultaneously, in large numbers of species. Eventually, systematists would code the morphologies or molecular structures of their subjects into matrices of discrete character states, choose the algorithm, and let the computer do the rest, at least up to the point at which it became necessary to choose between several almost equally plausible trees of relationship. Such innovations have led to big changes in the way in which paleontologists typically approach their material. In the process of becoming more rigorous, the science of paleontology has also become somehow more formulaic as intuition has been subordinated to procedure; but there is no doubt that the added clarity has led to enormous advances in knowledge, and even paleoanthropologists have begun to incorporate cladistic methods into their armamentarium.

UNFINISHED SYNTHESIS

The shift from authoritative pronouncement to cladistics heralded huge advances for the science of systematics, not least because conclusions now needed to be justified by something beyond mere bold declaration. But it was not the only revolutionary thing that was happening at the American Museum when I first arrived there, or perhaps even the most important one. For in its laboratories the very foundations of the New Evolutionary Synthesis, which by now formed the bedrock of our understanding of evolutionary biology, were under attack.

The Synthesis had given fossil species pretty short shrift, viewing them as ephemeral segments of evolving lineages that, if they didn't die out, gradually evolved themselves out of existence under the guidance of natural selection. This expected pattern of evolution left paleontologists with the pretty humdrum role of simply collecting fossils and using them to document those slow transformations. But by the time the 1970s came around, some paleontologists were finding it increasingly difficult to force the fossils they worked with into this framework of gradual change. Too often, the expected intergradations between earlier and later fossil species of the same group were failing to show up. Instead, what paleontologists were finding—and not for the first time—was that species tended to turn up quite suddenly in the fossil record, to linger around more or less as they were for extended periods of time, and then to disappear as abruptly as they had arrived.

Commonly, this failure of the fossil record to coincide with gradual-ist expectation had been explained away by the famous incompleteness of the fossil record: known fossils represent only the tiniest fraction of all the creatures that ever lived, and the expected intermediates simply hadn't been discovered yet. Darwin himself had used this excuse back in the middle of the nineteenth century, when he devoted a whole chapter of the *Origin of Species* to the deficiencies of the geological record. And yet, even then, some paleontologists had known that non-change was as much the message of the fossil record as change was. Back in the 1840s, no less than Hugh Falconer had remarked on the notable persistence of fossil taxa throughout the thick pile of rocks in India's Siwalik Hills, now known to cover some 4 million years of time.

In any event, by the early 1970s the Synthesis and the expectations of gradual change that flowed from it were ripe for reappraisal. This was duly supplied by Niles Eldredge, a young curator of fossil invertebrates at the American Museum. For his doctoral thesis at Columbia, Eldredge had studied a group of trilobites (ancient sea-bottom-dwelling invertebrates) that were found in rocks both in the Midwest and upper New York State. And he noticed, as Falconer had, that the basic picture among his organisms of interest was one of stability (stasis). Over a span of some 6 million years in the Midwest, only one single, sudden anatomical change suggested the replacement of a resident trilobite species by a close relative. He found the same succession, over the same period, at sites in New York. But with this difference: in one lone New York locality, both kinds of trilobite were represented. Rather than squeezing this information into a preconceived picture, Eldredge decided to trust the evidence of his eyes. Accordingly, he concluded he was witnessing the birth of a new trilobite species in that New York quarry.

The implication was that, after millions of years of business as usual, a swift event of speciation had taken place in New York, following which an environmental change had allowed the new species to spread to the Midwest, to displace its parent there, and to settle back into stasis. There was a certain drama in this story, but its message was certainly not the gradualist one preached by the Synthesis. Rather, an extended period of non-change had been interrupted by a short-term event of speciation, to be followed once more by stasis. What is more, just as Giambattista Brocchi had noticed in his Apennine marine shells a century and a half earlier, Eldredge's trilobite species had evidently had births, lifespans, and deaths. Species, far from being ephemera that evolved themselves out of existence, appeared to be real entities, with finite boundaries in time as well as space. Darwin's descendants had argued that speciation was a passive result of adaptive change under natural selection; but the new viewpoint made speciation the very basis of the change paleontologists saw in the fossil record.

A technical study of an obscure group of invertebrates was not guaranteed to garner much attention, but in 1972 Eldredge joined forces with his colleague Stephen Jay Gould to generalize his observations into a general theory of evolutionary pattern. The title of the book chapter in which they did this, "Punctuated Equilibria, an Alternative to Phylogenetic Gradualism," says it all. Eldredge and Gould argued provocatively that evolution, far from being the gradual process envisaged by Darwin, was typically episodic. Lineages routinely remain stable until they split, in a rapid process known as speciation. As we saw, Ernst Mayr himself had devoted a lot of thought to speciation, which in higher vertebrates like primates almost certainly depends on the cessation of physical contact between two parts of the same species. When such splitting occurs, it will almost certainly be as a result of external factors: maybe a shift in the course of a river, or the creation of an island by rising sea levels, or local climate change. The resulting isolation permits the species fragments to develop reproductive incompatibilities that ensure they will behave more or less as different entities should they ever find themselves reunited—although Mayr made the point that, to complete the process, "isolating mechanisms" might have to evolve as a result of wasted matings. Eldredge and Gould themselves suggested that most physical change in evolution probably occurs during "allopatric speciation" events of this kind, although it is notable that, as successful species spread, they tend to produce distinct local varieties.

The underlying message of "Punctuated Equilibria" was that, instead of representing deficiencies in the database, the "breaks" that paleontologists had always perceived in the fossil record might actually encode real information about evolutionary histories. And while Eldredge and Gould did acknowledge that their way of looking at things was just as likely as the Darwinian one to color phylogenetic interpretation, they were able to produce a wide variety of examples in which their new framework better fit the data than the Synthesis did. They also broached an apparent paradox that their new view raised. If the evolution of species is not somehow directional, as it necessarily is under the gradualist, selection-driven model, why do species belong to larger groups within which evolutionary trends over time are indeed apparent—as in the case of brain enlargement in the genus *Homo?* They found the answer in the nature of species themselves. As we saw early in this book, how species should be defined has been debated *ad nauseam.* But if species are not simply segments of lineages, and instead have births, lifespans, and deaths, then regardless of precise definition they can be seen as *individuals,* playing a role on the ecological stage that is analogous to the one the individual organism occupies under conventional selectionist theory. Just like individuals, species vary in their ability to survive and to give rise to offspring as they compete among themselves for resources in the environment. And triage at this level implies that, on the larger ecological scale, evolutionary trends can be generated through the differential survival and reproductive success of entire species.

Eldredge and Gould's reasoning started me thinking again about natural selection and the role it plays in Nature, and I have never felt the same way about selection since. The gradualist mind-set behind the Synthesis encouraged paleontologists to think of evolutionary trends in terms of particular characters, or at least of functional character complexes. Paleoanthropologists, including me, fretted about the selection-driven evolution of the human brain, or of the foot, or of the digestive apparatus, as if these things were separable from every other aspect of the organism and had discrete evolutionary trajectories that could be followed through time. With anatomical complexes abstracted in this way, evolution appeared as a process of optimization of traits, whereby populations became steadily better adapted to their environments over long stretches of time.

But think about that for a moment. And if you do, you'll notice that the reality is very different. Any feature you might want to follow in the fossil record is deeply embedded in a complex and autonomously functional organism that has to survive and compete in a multidimensional environment. Rarely will the fate of that organism depend entirely on the particular trait you're interested in. What is more, the nature of heredity is such that every gene plays a large variety of roles in the development of an organism, meaning that in most cases selection pressures cannot favor one particular aspect of a gene's activity without also adventitiously affecting its other functions. And this integration of the genome brings us abruptly back to the fact that the target of natural selection is not any specific feature, but rather the survival and reproductive success of the *entire* organism. Natural selection—which is, after all, no more than differential reproduction—simply cannot single out particular traits to favor. It can only vote up or down on the whole package. You succeed or you fail in the reproductive stakes as the sum total of everything you are. So while it might seem an evident advantage to be, for example, the smartest member of your population, this distinction is unlikely to assure your overall success if you are also the slowest, or the shortest-sighted, or even simply the unluckiest of your compatriots. Similarly, in evolutionary terms it is hardly advantageous to be the most magnificent exemplar of your species if, in the larger scheme of things, that entire species is being outcompeted into extinction, or eliminated by environmental change. The bottom line is simple: individuals are such complex genomic and physical packages that within-population processes will rarely if ever be about optimization in specific features.

None of this means that natural selection is not important; indeed, it is a mathematical certainty that selection will be operating in any population in which more individuals are born than live to reproduce successfully. In other words, selection operates in *every* population. But, most of the time, selection will be acting to keep things the way they are, by trimming off the unwieldy extremes at both ends of the spectrum of variability. This is the long-recognized phenomenon known as stabilizing selection. Further, it is often difficult to see selection acting as a pervasive and persistent agent of long-term change in a world in which—as is becoming increasingly clear—environmental fluctuations are the rule rather than the exception. Usually the best strategy is not to put all your eggs into the same evolutionary basket, and it is no accident that extinction rates are higher among specialist species than they are among generalists. This is not, of course, to say that instances do not exist in which organisms have evidently responded quite strongly and rapidly to modifying pressures. For example, it is probably not coincidental that male chimpanzees, which constantly compete for access to large numbers of females, have very much larger (hence more productive) testes than those of the male gorillas that control individual females for extended periods. But the majority of such cases involve attributes that are very close to the reproductive process itself, or that are particularly critical to day-to-day survival (think of persisting lactose tolerance in adult members of cattle-raising human populations). And most attributes are simply not this specific. Being the best is often no more effective in the reproductive stakes than just being good enough to get by.

If you allow considerations like these to draw you away from the traditional focus on competition between individuals, and you can instead begin to see species, or at least populations, as important actors in the evolutionary drama, your perspective on the entire process changes. Gradual change begins to look significantly less plausible as a routine phenomenon, and it is possible to see why species tend to remain stable over time. Ironically, the very first thing I was taught in my very first, Synthesis-infused evolution course was not the mechanisms of change. Instead, I learned about the so-called Hardy-Weinberg Equilibrium. This mathematical formulation states that in the absence of extraneous influences, gene frequencies in a population will resist change from generation to generation. Evolution, if you will, represents a disruption of that normal equilibrium. And such disruption is far less likely to be precipitated by some dynamic that is internal to a population than by some kind of shock from the outside, such as those routinely delivered to our human precursors over the Pliocene and Pleistocene epochs during which Hominidae evolved. Climates in those times were notoriously unstable, which meant that hominids were living in unpredictable and constantly changing conditions. To cope with the resultant oscillations in the resources available to them, it was crucial for hominids to stay adaptable and not to specialize too much on exploiting any one kind of environment. Not only is natural selection far too gradual a process to track such changes, but constant fine-tuning of hominids to their habitats would actually have been a huge disadvantage to them.

At the same time, though, it is undeniable that those environmental fluctuations would have had a brutal effect on hominid populations. In hard times those populations would have been fragmented into small isolates that hung on in particularly favorable places while intervening areas became hostile to human occupation. When conditions improved, these small remnants could expand once more from their refugia, bringing them back into contact with their former peers. This repeated process was doubt-less hard on the individual hominids involved, but it provided exceptional opportunities for evolution. Very small populations are much more geneti-cally unstable than large ones, and they provide optimal conditions for incorporating the randomly arising genetic novelties that furnish the very basis of evolutionary change. What is more, as Mayr himself had pointed out, tiny populations also supply the circumstances in which speciation (which, as I later realized, is not simply a passive consequence of morpho-logical change) is most likely to take place. That is one part of the equa-tion, but equally important is the other. Depending on what might have happened during the period of separation, one of two things might occur when contact was reestablished: If speciation or something close to it had occurred in isolation—if something had happened to disrupt reproductive compatibility—the result would be competition between the populations, resulting in the elimination or displacement of the competitively inferior. If, in contrast, any differentiation that had occurred had not involved genetic incompatibility, genetic novelties from both sides could be incorporated into the recombined population. Either scenario would result in change.

All of this pointed toward an evolutionary process—or, more strictly speaking, the whole slew of different processes that we can see in retrospect as contributing to evolutionary change—that was much more complex and multidimensional than the slow adaptive modification envisioned by the Synthesis. And it had huge implications for how evolutionary biologists should view all aspects of the history of life on Earth, including that of the hominid family. But, mesmerized as they were by the gradualist viewpoint, many evolutionary biologists initially had great difficulty with the new way of seeing things. And in consequence, Eldredge and Gould immediately found themselves—entirely unfairly—accused of such heresies as salta-tionism, or denial of adaptation, or worse. This was nothing new: Darwin's publication of the *Origin of Species* had been followed by a brief period of widespread shock and horror before the basics of evolution became widely accepted within a couple of decades. In just the same way, the years fol-lowing Eldredge and Gould's publication witnessed the reasonably rapid acceptance of their views, and widespread acknowledgment that numer-ous factors entirely unrelated to excellence of adaptation had played crucial roles in determining the history of life.

The new perspective helped evolutionary biologists to understand that the evolutionary histories of successful groups of organisms tend as much to reflect a process of experimentation, with the potential inherent in their founding adaptations, as to embody the results of a slow and steady process of refinement. Still, largely as a result of their one-species focus, paleo-anthropologists have proven rather slower than most to take this lesson on board.

CHAPTER 6

THE GILDED AGE

ONE OTHER REASON WHY PALEOANTHROPOLOGISTS WERE
slow to pick up on changes in the wider world of systematics and evolutionary biology may have been that during the late 1960s and early 1970s they
were already having trouble digesting a wealth of new additions to the human fossil record. According to legend, it all started with a state visit paid by
the Ethiopian emperor Haile Selassie to the newly independent Republic of
Kenya in 1966. One of the people he met during this visit was Louis Leakey,
who proudly showed off some of his new Olduvai hominids. When Selassie
asked why Tanzania should have wonderful human fossils while Ethiopia
didn't, Leakey told him that nobody had looked—which was functionally
if not literally true. Shortly thereafter, he found himself with an imperial
invitation to rectify the situation.

With his hands full promoting both his own fossils and the researches
of a succession of glamorous young primatologists, the aging Leakey passed
the principal responsibility for organizing fieldwork in Ethiopia to his re-
spected American colleague Clark Howell. The area of the country he had
his eye on was the lower basin of the Omo River, which emptied into the
almost entirely Kenyan Lake Turkana (then called Lake Rudolf) just north
of the Ethiopian frontier. Since the French paleontologist Camille Arambourg also had a claim on the Omo, having prospected there back in the
1930s, a Kenyan-American-French team was eventually assembled to do the
work. By then approaching the end of his career, Arambourg nominated

his younger colleague Yves Coppens to represent France, while Leakey delegated his Kenyan responsibilities to his middle son, Richard. At the time, Richard was making his living as a safari outfitter and hunting guide, and his sole qualifications for the Omo job consisted of his experience as a paleontology brat and his acknowledged organizational skills.

The international expedition reached the field in 1967, with each national team working in its own concession area. From the very beginning, Howell thought expansively. To carry out the Omo fieldwork he assembled a large team of specialists: geochronologists to do the dating, stratigraphers to work out the succession of rocks, paleontologists to study the fossils, taphonomists to figure out how they had been formed and preserved, archaeologists to analyze the stone tools, and so on. This was the birth of Big Paleoanthropology, and Howell's approach has served as a model for large-scale fossil-hominid–oriented fieldwork ever since.

In the event, Howell's field area turned out not to yield much in the way of well-preserved hominid fossils. But it did contain a great thickness of mostly river-laid fossiliferous sediments that were regularly interspersed with lava flows and tuffs (volcanic ashfalls) that were datable by potassium-argon (K/Ar). Because the area was geologically unstable, a lot of faulting had occurred, which made the succession of rocks difficult to figure out. But, significantly, this same fracturing also exposed rocks—and the fossils they contained—from throughout the time period between about 1 and 3 million years ago. In the end, Howell's expeditions were able to document a finely calibrated sequence of faunas that allowed sites to be placed in the succession even when they lacked good chronometric dates. In this regard the fossil pigs, as studied by the paleontologist Basil Cooke, proved to be particularly useful: if you knew the species of fossil pig present, you had a pretty good idea how old a site was.

Because the Omo sediments had typically been laid down by relatively fast-moving waters unfavorable to fossil preservation, most of the hominid fossils found by Howell's team were pretty fragmentary. Indeed, most of them were just individual teeth. But, when he finally analyzed his haul, Howell was able to conclude that between about 1 and 2 million years ago he had mainly fossils of a hominid with huge, flat chewing teeth. These he assigned to *Australopithecus boisei*, the name by then commonly used for Louis Leakey's robust hominid from Olduvai. Howell compared most teeth from between about 2 and 3 million years ago to those of *A. africanus* from

South Africa, and he assigned teeth from around 1.85 million years ago to *Homo habilis*. Finally, he allocated some smaller teeth, about 1.1 million years old, to *H. erectus*. There was something here for everybody.

For the same reasons, hominid fossils from the French concession area were similarly fragmentary. But one that ultimately attracted attention was a toothless lower jaw, dated to about 2.6 million years ago. This mandible was not itself enormous, but it had clearly once contained very large chewing teeth. Arambourg and Coppens decided that this mysterious fossil did not belong to any species yet described, and with fine French abandon (the French were very slow to adopt the tenets of the Synthesis) called it *Paraustralopithecus aethiopicus*.

Meanwhile, Richard Leakey had been chafing. He was the most junior team leader, and he and his Kenyan contingent had been assigned the youngest and least promising of the Omo sediments for prospection. Still, in 1967 his group did recover two partial skulls in the Kibish region, one modern-looking, the other somewhat less so. These were thought at the time to be around 125,000 years old, though they are now dated closer to 200,000 years ago and are recognized as critical evidence for understanding the emergence of *Homo sapiens*. But the Kenya contingent found no early hominids of the kind that had made the senior Leakeys' names, and certainly nothing that could be identified as the "earliest *Homo*," the Holy Grail of the Leakey family. So Richard, who has difficulty being a junior partner in anything, borrowed a helicopter that had been rented by the expedition. He flew it south into his homeland of Kenya where, on a previous supply trip, his interest had been piqued from the air by the sedimentary rocks that lined the eastern shore of the exceedingly remote Lake Turkana. On landing, he saw fossils eroding out from the sandstones around him, and immediately resolved that it was there, on his home turf, that he would henceforth expend his energies. He informed the National Geographic Society (and his reluctant father) that the funds it was contributing to the Omo expedition would be better spent in Kenya and, after secretly writing to the Ethiopian government to denounce his American partners for incompetence, led his first survey group to East Rudolf in 1968. Bankrolled by the National Geographic Society, this survey confirmed the area's potential, and in 1969 fossil collecting started there in earnest.

Almost immediately, a splendidly preserved hominid cranium was found that reminded its finders of the Zinjanthropus skull from Olduvai, although its face was significantly shallower from top to bottom. It no longer contained any teeth, but its dental proportions had clearly been very much like those of the Olduvai form, with greatly reduced front teeth and hugely expanded chewing ones. The skull was assigned to *Australopithecus boisei* and received the memorable name of KNM-ER 406 (for "Kenya National Museum-East Rudolf 406," its catalog number). ER 407, a second cranium found in the same year, was much more lightly built, and was reported by Richard as a potential member of the genus *Homo*. Also found were some simple stone tools comparable to those found in the lowest levels of Olduvai. These had been found within a tuff (known as the KBS) that had been erroneously K/Ar dated to 2.6 million years ago (the real date is 1.95 million, as initially suggested by the fossil pigs found below it), and Richard felt it likely they had been made by hominids that resembled ER 407. In short, the situation at East Turkana seemed from the very beginning to have closely resembled what the elder Leakeys had been reporting from Tanzania, with an absolute separation between *Australopithecus* and *Homo* lineages very early in time. The basis for this assessment seems to have been little more than family proclivity: Richard fought to the bitter end for that inaccurate 2.6-million-year date for the KBS tuff, and ER 407 itself was later re-identified as a lightly built female *A. boisei*. On another level, Richard actually seems never to have had any interest in systematics (a lack he shared with most of the anatomists he conscripted to describe his hominids), and he long insisted that when fossils assigned to *Homo* were published in the literature, they should appear as "*Homo* sp.," meaning that their species remained undetermined.

Still, none of this made continuing discoveries at East Turkana any less exciting, even if one find after another appeared to support very early establishment of separate *Australopithecus* and *Homo* lineages. Each field season brought many new hominid fossils, mostly fragmentary but inevitably with something unexpected among them. In 1970 a 1.7-million-year-old partial cranium (ER 732) was found that was by general consent a female *Australopithecus boisei*. It was more lightly built than the penecontemporaneous 406, and it lacked the large crest in the midline of the braincase to which the enormous chewing muscles of the latter had attached. But all agreed that it shared the same basic morphology. And in doing so, this cranium undermined the remaining adherents of the Single-Species Hypothesis—by then led by C. Loring Brace's Ann Arbor colleague Milford Wolpoff—who

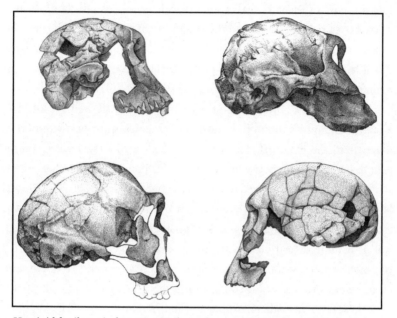

Hominid fossil crania from East Turkana discussed in this chapter. Clockwise from top left: KNM-ER 732, identified as a female Paranthropus boisei; *KNM-ER 406, male* P. boisei; *KNM-ER 1470, early on believed to substantiate* Homo habilis *as a taxon but now often placed in* Kenyanthropus rudolfensis; *KNM-ER 3733, first identified as an "early* Homo erectus*" but now often assigned to* Homo ergaster. *To scale. Drawn by Don McGranaghan.*

had continued to argue that the highly distinctive robust and gracile South African hominids were nonetheless males and females of the same species. Clearly, multiple lineages were involved. Years later, the plot thickened still further when a 1.9-million-year-old frontal bone (ER 23000) was discovered that matched its counterpart structure in the Olduvai Zinjanthropus skull—but not that in the ER 406 skull. But nobody has yet seemed very anxious to come to grips with this evident complication.

The next year's hominid trove from East Turkana included a 1.4-million-year-old lower jaw (ER 992) that, mainly because of its fairly modest dental dimensions, reminded many in the Turkana group of *Homo erectus* from eastern Asia, though it was announced merely as a member of *Homo*. In 1975 the zoologists Colin Groves and Vratislav Mazak made this distinctive mandible the type specimen of a new species, *H. ergaster* ("work man," for all the stone tools found in the same sediments). This predictably raised the ire of Richard Leakey and his colleagues, who disdainfully dismissed both the new name and the credibility of the two unauthorized interlopers who had come up with it. But Groves and Mazak's bold innovation did at least open the door on the principle that some diversity might exist among the *Homo* fossils of the Turkana Basin.

During the 1972 field season the real newsmaker showed up: ER 1470, a hominid cranium found in hundreds of fragments just below (and thus older than) the KBS tuff, which was still believed by the Leakey contingent to be somewhere between 2.6 and 2.9 million years old. When reconstructed, its braincase proved to be remarkably capacious: 800 ml. This was later reduced to 750 ml, but even then it was significantly larger than the estimated 640 ml of Louis Leakey's original *Homo habilis* from Olduvai. It was difficult to determine exactly how the face had originally been hafted on to the cranial vault. But as reconstructed it appeared pretty flat, and clearly it had originally housed fairly large teeth, arrayed in a broad palate.

These features made 1470 quite unusual, but unfortunately its fragmentary condition meant the specimen lacked a lot of anatomical detail, which made—and still makes—it very difficult to interpret. Mainly because of its large brain Richard assigned his new specimen to the genus *Homo,* but believing it to be well over 2.6 million years old, he was reluctant to allocate it to the same species as his father's smaller-brained *Homo habilis* from Olduvai. As usual, he plumped for *Homo* sp. It is ironic, then, that 1470 was ultimately what persuaded many paleoanthropologists that *H. habilis* was indeed a real independent entity.

Meanwhile, the discoveries kept pouring in. In 1973 two more crania were found. ER 1805 is quite fragmentary, and still resists interpretation. But the lightly built ER 1813 is nicely preserved; and although Richard Leakey's initial comparison was with South African *Australopithecus africanus*—its brain volume of not much more than 500 ml puts it in the right range—Clark Howell eventually opined that it actually represented a female *Homo habilis.* Many have adopted this identification, although their acceptance may mainly emphasize just how useful it is to have a "wastebasket" taxon around into which you can sweep the untidy odds and ends that you do not otherwise know what to do with. Predictably enough, Richard never liked this development; he eventually used the resemblances he saw between 1813 and some of the Olduvai Bed II materials to argue that neither belonged to *H. habilis.* In the meantime, he also found himself forced

to concede on the age of the KBS tuff. This concession made 1470 almost
exactly the same age as Olduvai Bed I *H. habilis*, and thus more likely to be
the same thing. So, despite the miscellaneous nature of the fossils that had
been swept into it, *H. habilis* had finally, and largely by default, become an
entity to be reckoned with.

East Turkana had yet more surprises in store. The 1974–75 field season
produced ER 3733, a 1.78-million-year-old cranium (then thought to be between 1.3 and 1.6 million years old) that looked like nothing else yet found
in Kenya. Together with his anatomist colleague Alan Walker, Richard
Leakey assigned this specimen to *Homo erectus*. They reported a particular resemblance to the Chinese Zhoukoudian hominids, and pointed out
that the discovery administered the final coup de grâce to the still-lingering
Single-Species Hypothesis. But while Leakey and Walker clearly took pleasure in pointing out that the demise of the unilinear model meant a new
schema for human evolution was needed, they declined to produce one of
their own. The new skull had a brain volume of 848 ml, on the smallish side
for *H. erectus* but within hailing distance; other than its general proportions, though, it did not have many of the hallmarks of the Chinese material
to which it had been compared. Not long afterward a 1.6-million-year-old
partial cranium, ER 3883, was discovered and also attributed to *H. erectus*. While it, too, lacked the anatomical specializations of the eastern Asian
hominid, it didn't look exactly like 3733, either. That its differences from
the latter weren't an anomaly was shown by a smaller and slightly older
skull fragment, ER 3732, that is identical in comparable features.

THE JUNIOR LEAGUE

The scientist recruited by Richard Leakey to lead research in archaeology at
East Turkana was the South African Glynn Isaac, who joined the project in
1970 after working extensively with Mary Leakey at Olduvai. Isaac shared
her interest not only in the stone tools that early hominids had made, but
also in the nature of the sites at which those tools were found. Typically, at
Olduvai these consisted of lakeside scatterings of stone tools and the debris
created in the process of producing them, plus an array of animal bones.
Mary believed that such sites were favored points on the landscape, to which
hominids had repeatedly brought carcasses for butchery, and consequently
referred to them as "living sites."

At East Turkana a similar situation seemed to prevail, and Isaac rapidly developed a model of early hominid behavior that viewed those "campsites" as focal points in a lifestyle that involved hunting and scavenging in the surrounding territory. This lifeway would also have entailed bringing in carcasses for butchery, using tools made from rock types that were not always available in the immediate vicinity, and Isaac thought it would have necessitated a central sharing system of the food thus obtained. A further assumption was that some division of labor existed among females and males: the females, burdened by offspring, would have been limited to foraging in the vicinity of the "home base," while the males ranged more widely in the search for carcasses and hunting opportunities. Social features of this kind seemed to indicate considerable sophistication in early hominid societies, which would perforce have displayed complex systems of communication and cooperation and the development of intricate relationships among individuals. Bipedality, social reciprocity, and even some rudimentary form of language were thus bound together into a remarkably humanlike package. This perspective fed back nicely into the then popular "Man the Hunter" view of human origins; and it also extended the reconstruction of hominid history into the behavioral domain, by projecting the image of *Homo sapiens* back into the past.

Such constructions of early hominid life did not go unopposed. In the mid-1970s the American archaeologist Lewis Binford began to point out that it was perhaps unwise to try to see early hominids as simply junior-league versions of ourselves. Maybe they had their own way of doing business. And archaeologists had a better chance of finding out what that way of business was if they paid more attention to the facts themselves, and less to speculation about them. As a thoughtful and meticulous scientist, Isaac was sensitive to such strictures, and eventually he began a reexamination of the East Turkana evidence. The initial goal was to find out whether the associations between the bones and stone tools at the archaeological sites were indeed due to human activities, and if so, just what those activities were. And if they proved not to be, what had caused the associations? By the time he died in 1985, at the tragically early age of 47, Isaac had made considerable progress on these questions. Most importantly, he had shown that by around 2 million years ago the landscape at East Turkana definitely bore the hallmarks of human activity. Thus the cut marks seen on many of the animal bones had been made by the sharp edges of stone tools, while a high proportion of those sharp edges bore wear of the kind that is made by cutting fresh meat and bone. Those slashing incisions are indeed butchery marks and, as Isaac drolly observed, "one can only presume that [the hominids] ate the meat that they cut." What is more, certain body parts were preferentially represented among the cut-marked bones, suggesting that those bits of the carcasses had been carried in to the central spot after they had been initially obtained elsewhere. Some suggestion of the "home base" thus remained, although Isaac began to use the term "central-place foraging" rather than "food-sharing" to describe the activity involved. The resulting notion of early hominid behaviors was both more shadowy and less dramatic than the one it replaced. But as Isaac himself put it, this paring down was necessary "to avoid . . . creating our origins in our own image."

LUCY AND THE FIRST FAMILY

Richard Leakey was not the only spin-off beneficiary of Clark Howell's pioneering African fieldwork. In 1970 and 1971 Howell's graduate student Donald Johanson had accompanied him to Ethiopia. There, through members of the French Omo team, Johanson met Maurice Taieb, a geologist doing research for his doctorate in the Afar Triangle, an area of inhospitable desert in the far north of Ethiopia. The triangle lies where the East African Rift Valley—the great geological scar that bisects Africa all the way south to Mozambique—meets the Red Sea and the Gulf of Aden. The Rift Valley has been of huge importance to paleoanthropology, for it is the structure within which the fossil paradises of Omo, Turkana, and Olduvai all lie. In each of these places, the sediments containing the hominid fossils had been eroded from the rising highlands at the valley's sides. During his first survey in the Afar, Taieb had noticed fossils eroding abundantly out of what he thought were Plio-Pleistocene sediments, and he asked Johanson if he might be interested in looking at them. No second suggestion was needed, and early in 1972 Johanson joined Taieb, Coppens, and Jon Kalb for a survey of the region. Kalb was an American geologist resident in Addis Ababa who had already visited the Afar with Taieb. The team found a paleontologist's wonderland at a place called Hadar, an area of harsh badlands through which the broad River Awash meandered, supporting a thin strip of gallery forest. To judge by the Omo faunal yardstick, the fossils the quartet found eroding in abundance out of the sediments all around them seemed to be about 3 million years old. Fieldwork at this extraordinary place started in earnest the next year.

In contrast to Turkana, where the standout finds were those many skulls, it was postcranial elements—bones of the limb skeleton—that were to be the biggest stars of the show at Hadar. And even just in prospect, this was exciting. For if the preliminary age assessment was right, the sediments at Hadar were older than anything that had yet yielded hominids. They were old enough, indeed, to yield crucial information about the origins of bipedality, which all by now agreed had been the basic hominid acquisition that set our hominid family off on its fateful and unprecedented course. Expectations were high, and that first full field season in 1973 did not disappoint. A huge collection of fossil mammals was made, among them the distal (far) end of a femur and the proximal (near) end of a tibia. Together, these composed the knee joint of a rather small hominid individual. This might not sound like much; but the knee, as it happens, is hugely diagnostic of bipedality. In a quadruped the weight is transmitted straight down from the hip joint to the ground, so that when it moves each leg stabilizes one corner of the torso, very much as the legs of a square table do. But when an animal that is basically adapted to four-legged movement—a chimpanzee, say—stands upright and moves bipedally, everything changes. With every step it pivots its whole body around the supporting foot, swinging each leg forward in a wide arc, and moving the body's center of gravity clumsily from side to side. This wastes a lot of energy. The femur of a bipedal hominid avoids this inefficiency by angling inward from the hip. The knees and feet pass close together during walking, and the center of gravity progresses in a straight line. All this is made possible by the formation of an angle between the shaft of the femur and the joint surface below it, which makes for a very distinctive bony structure that, tellingly, was present in the Hadar knee. At a hastily arranged press conference in Addis Ababa, Johanson announced that at some time between about 3 and 4 million years ago, a bipedal hominid had already walked erect in Ethiopia.

After this, the finds at Hadar came thick and fast. In 1974 some hominid jaws were found, followed by "Lucy," to this day the most famous hominid fossil ever discovered. What exactly it was about Lucy that so caught the public's imagination is still hard to say, although the snappy name, much more imagination-grabbing than a dry museum catalog number, clearly had a lot to do with it. So also did Johanson's unflagging promotion of his remarkable find, as he evoked the poignant image of a young female, almost human but not quite, who had died all alone in the Ethiopian bush some 3.2 million years ago. But Lucy's fame was also underpinned by the fact that here, for the first time, was a truly ancient hominid fossil that preserved enough for observers to be able to form a mental image of the entire living, breathing individual. Lucy's tiny fossil skeleton (she had only stood about three and a half feet tall) was certainly incomplete (in fact, her skull consisted only of her lower jaw and some fragments of vault), but enough of the bones of her body remained to reveal a lot about the sort of creature she had been in life. And those fossils clinched the claims Johanson had made at that first press conference in Addis. From the waist down, at least, Lucy's bones were unquestionably those of an upright walker. In contrast to the deep, narrow pelvises of the quadrupedal apes, her pelvis was wide and shallow to support the viscera that now lay above it, rather than in front. That pelvis was, indeed, basically like our own, although it flared even more widely. Lucy's thighs were fairly short, but they had angled inward from her widely spaced hip joints, and her ankle had been constructed to transmit her weight straight downward to her unfortunately missing feet.

So who, exactly, was Lucy? Just before she was discovered, but after a couple of hominid jaws had already come to light, Johanson was visited at Hadar by Richard Leakey. Recognizing that the jaws were clearly not those of the massive *Australopithecus boisei*, Richard predictably enough proposed that they must represent an early species of *Homo*. Johanson was already entertaining this notion himself, so when he, Taieb, Coppens, Kalb, and the fossil pollen expert Raymonde Bonnefille described the 1973 and 1974 Hadar hominids, they suggested that the jaws might be assigned to a member of the genus *Homo*. But at the same time, they compared Lucy and the knee joint to South African *A. africanus* (of which several postcranial bones were by then known), and did not commit themselves to a precise identity. The plot thickened in 1975, by which time Ethiopia was in the throes of revolutionary turmoil. Despite political disruptions the team returned to Hadar, coming away with another haul unprecedented in paleoanthropology: a trove of over 200 fragmentary hominid fossils, derived from a dozen or more individuals who seemed to have perished together in some kind of catastrophic event, possibly a flash flood.

What was most remarkable about the "First Family," as this assemblage of fossils was dubbed, was its large size range. If all of the individuals

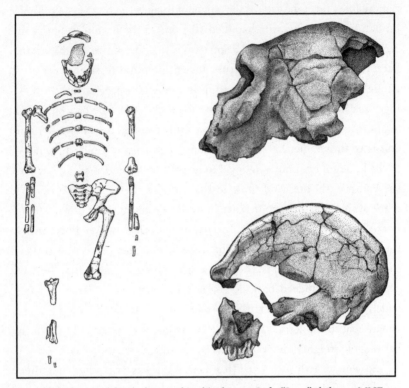

East African hominid fossils discussed in this chapter. Left: "Lucy" skeleton, MNE AL 288, from Hadar, Ethiopia, assigned to Australopithecus afarensis. *Upper right: Bodo cranium from Middle Awash, Ethiopia, first mooted as a* Homo erectus *but now often seen as* Homo heidelbergensis. *Lower right: comparatively recent partial skull LH 18, from Laetoli, Tanzania. Crania to scale. Drawn by Don McGranaghan.*

represented had belonged to the same species—as would certainly have been the case if all had belonged to a single social group—this disparity suggested that the species concerned was strongly sexually dimorphic, meaning males were much larger than females. Of course, modern human—and chimpanzee—males do tend to be larger than females, but the difference at Hadar was much greater, on a par with what is seen today only in the apparently very specialized gorillas. This was puzzling; but what was especially gratifying about the First Family collection was that it included elements of the body—most importantly, of the hands and feet—that had not been preserved in the Lucy skeleton. The team returned to Hadar once again in

1976, finding some more jaws and discovering the first stone tools known from the area, at a site dated to about 2.5 million years ago. And at least for many years that was it for more fossils, because further dire political events in Addis Ababa ended fieldwork for the next decade and more.

THE MIDDLE AWASH AND LAETOLI

Meanwhile, many other things had also been happening in Ethiopia, entertainingly chronicled by Jon Kalb in his 2001 book *Adventures in the Bone Trade,* mandatory reading (along with Eustace Gitonga and Martin Pickford's *Richard Leakey: Master of Deceit*) for anyone interested in the sociopolitical dynamics of the Gilded Age of paleoanthropology in eastern Africa. To cut a long story short, after much internecine intrigue the Afar team splintered in late 1974. Johanson and his French colleagues remained at Hadar, while Kalb shifted his attentions a little way upriver, to an area of badlands known as the Middle Awash. In 1976 his team discovered a hominid cranium, now known to be about 600,000 years old, at the Middle Awash site of Bodo. At 1,250 ml, its brain volume was fairly respectable, and the closest morphological comparison seemed to be with the Broken Hill (now Kabwe) skull from Zambia that Arthur Smith Woodward had assigned to *Homo rhodesiensis.* Diplomatically, its discoverers reported that the Bodo cranium was "less archaic" than Asian *H. erectus,* but "more archaic" than the quite recent materials Richard Leakey had reported from Omo Kibish. The Middle Awash, where sporadically exposed fossil-bearing rocks document the last 6 million years, eventually came to loom large in the history of paleoanthropology. But back in the 1970s the discoveries most germane to the interpretation of the Hadar fossils were being made far to the south, at Laetoli in Tanzania.

Laetoli, not very far from Olduvai, had first been visited by the elder Leakeys back in the 1930s, when they had recovered a primate lower canine tooth. At the time they thought it was a monkey's, but it later turned out to have belonged to an early hominid. A few years later, the German explorer Ludwig Kohl-Larsen found a fragment of lower jaw not far away. In 1950 another German, Hans Weinert, concluded that this mandible belonged to Franz Weidenreich's Javan genus *Meganthropus,* and created the new species *Meganthropus africanus* for it. There things rested until Mary

Leakey returned to Laetoli in 1974. Over the next eight years Mary's team discovered some 30 early hominid fossils, ranging from isolated teeth to two quite complete lower jaws (LH—for Laetoli Hominid—2 and 4) that were between 3.6 and 3.8 million years old. It also found a fairly complete skull in the much later Ngaloba Beds that are also exposed in the area. This specimen (LH 18) came in with a cranial volume of around 1,200 ml, and the two anatomists who described it rather noncommittally remarked on its general resemblances to other "archaic *Homo sapiens*" crania from Africa.

The real jewels of the Laetoli fieldwork, though, were fossils of another kind altogether: trackways preserved in ancient ashfalls. Many mammals, from giraffes to African hares, had left their footprints at Laetoli in muddy ash that had been spewed out by a nearby volcano, then dampened by rain (in some places you could see the marks of individual raindrops). Those prints had then dried hard in the African sun, soon to be covered by more ashy sediment until exposed again by erosion, millions of years later.

Among those mammals was a pair of ancient hominids who had walked across the plain at Laetoli some 3.6 million years ago, leaving behind them the most famous footprints in history. In an 88-foot-long section of tuff, some 70 hominid footprints are preserved in two parallel trackways that head toward Olduvai in an unwavering straight line. The prints are those of undoubted bipeds who had walked unhesitatingly on two feet. One of the trackways, made by a larger individual, is rather blurry, and it has been suggested that a third and even a fourth hominid had followed, treading in the prints. But the other trail has some very sharp impressions, and the best of its prints show clear evidence for a strong heel strike and toe-off, the hallmark of modern walking. To some degree at least, the feet seem to have had longitudinal arches, and the great toes were certainly in line with the others. Certainly, they were not divergent like those in the grasping feet of apes. In sum, the Laetoli prints do not look exactly like those made by the feet of modern people, but whoever had made them was already confidently up on two feet.

That much is certain. But from the point of view of your author, a museum curator charged in 1990 with re-creating the Laetoli scene in a diorama, there was something rather odd about the trackways. The two individuals who made them were clearly of very different size, and would normally have had different stride lengths. Yet as Peter Jones, one of their excavators, pointed out to me, their footprints are matched, foot for foot, over their entire preserved distance. To achieve this, the two hominids must have been walking in step, side by side. What is more, the tracks are so close to each other that if so, their bodies must have been in contact, perhaps even awkwardly. What were they doing? Were they walking hand in hand? Linking arms? Were they carrying something between them? Anything is possible: there really is no way to choose objectively between the alternatives. Yet a diorama is necessarily a very explicit statement, because to be convincing the illusion of nature has to be complete. While we didn't want to go any further beyond what the evidence directly supported than we absolutely had to, there was no way to equivocate. Since our vignette of ancient hominid life had to be specific about what our subjects were doing, we sought the least loaded option. Other than small size we had no basis to conclude that the smaller individual was a juvenile, and we felt that to portray it as one would have been an unacceptable leap. So, on what seemed to be the reasonable assumption of high sexual dimorphism, we showed a large male with one arm draped over the shoulder of a smaller female. Both are shown looking around rather apprehensively as they trudge across a dangerous open landscape, teeming with predators, toward the forested shelter of the Olduvai Basin. I like to think they were taking comfort from each other. Still, we inevitably found ourselves excoriated by feminists for our paternalistic representation.

But whatever exactly they were doing, *someone* had made those trackways at Laetoli, some 400,000 years before Lucy was born. Who were they? The natural assumption was to implicate the hominids whose fossils had been found in penecontemporaneous sediments nearby. But for a while at least, the identity of these creatures remained something of a guessing game. Mary Leakey had by then adopted her son Richard's policy of discouraging everyone who worked with her from making systematic judgments about any fossils collected under the Leakey aegis. Of course, those fossils did have to be described and put on record, and this mundane task was entrusted to Tim White, a student of Milford Wolpoff's who had come to work at Laetoli after falling out with Richard at East Turkana over the age of the KBS tuff. White meticulously carried out his remit in one of the most unrelentingly tedious papers ever written by a paleoanthropologist. Hewing literally to the established Leakey policy, he said not one word in it about what those Laetoli hominids might have been, or even about what they might most usefully have been compared to.

AUSTRALOPITHECUS AFARENSIS

In the summer of 1977 Don Johanson, by then a curator at the Cleveland Museum of Natural History, asked White to bring casts of the Laetoli fossils to Cleveland so they could be compared to the Hadar materials, the only other hominids known at the time from the 3- to 4-million-year period. It wasn't clear at the outset that White and Johanson were going to agree on anything, especially since White had trained under the guru of the Single-Species Hypothesis, while Johanson's belief was already on record that more than one species might be represented at Hadar. One obvious complication was the large size range of the Hadar hominids, but there were also morphological issues. For example, the tiny Lucy had rather small lower front teeth, resulting in a distinctively V-shaped lower jaw, while larger Hadar mandibles had relatively bigger front teeth, and blunter curves to the jaw.

Nonetheless, while by Johanson's own account they did indeed start from very different positions, the two young researchers soon converged on White's single-species preference. First, they agreed that all of the Hadar and Laetoli fossils could be distinguished from all other known apes and hominids. From that common point they also decided that, while there were huge size disparities in the Hadar sample, there was a gradation between the extremes with no obvious breaks. Finally, they settled on the conclusion that most of the shape differences within the combined sample were allometric (consequent on the size differences), while the rest were merely of the kind that naturally occur among individuals of the same population. Presto! In the period between about 3 and 3.7 million years ago, a single distinctive and very sexually dimorphic hominid species had been present both at Hadar and at Laetoli.

Having satisfied themselves that they had before them the bones of a single hominid species, Johanson and White proceeded to put together a portrait of its members. These had been well adapted to bipedality, and although even the biggest males had not stood much taller than about four and a half feet, they had been substantially bigger than the females. To judge by the markings their muscles left on their bones, both sexes had been powerfully built. The legs of these hominids were relatively short compared to their arms, and their basically humanlike hands were a bit longer and more curved than ours today. Their torsos tapered upward, narrowing the shoulders. Their large faces had supported jaws that converged toward the front, forming a dental arcade that was neither parabolic like ours, nor parallel-sided like an ape's. Their canine teeth resembled those of apes in having a rather conical shape, but approached those of humans in their smallish size. Their incisors were quite big, and their molars, though large, were nothing like the great flat expanses of the robust australopiths. In contrast, their braincases were small, having sheltered brains in the chimpanzee volume range. They predated the most ancient stone tools known from Hadar, and had not been toolmakers.

So Johanson and White had before them a distinctive early hominid species. But how should they classify it? Largely on the basis of its great age—and not on its morphology, as a systematist would insist—they decided that their new species was the "stem" hominid, ancestral to all later members of the family including both *Australopithecus* and *Homo*. Logically, if their new form was truly the ancestor of both of these genera, it belonged to neither one, meaning that it would have made sense to create a new genus for it. But in post-Mayr paleoanthropology any such move was unthinkable. So instead, the two paleoanthropologists settled conservatively for the genus *Australopithecus*. This left the issue of species. Since they had already decided that their new form was unlike anything else known, they needed a new species name. In this regard, they made a rashly bold move. Every new species needs a "holotype" specimen that bears its name and serves as the standard to which every other putative member has to be compared. Since their fossils came from two localities that were geographically far apart, Johanson and White wished somehow to bind them together. To achieve this symbolic unity, they gave the name *Australopithecus afarensis* to the combination, after the location from which it was most abundantly known—but they designated as holotype a Laetoli mandible, LH 4.

This not only violated taxonomic convention; it tempted fate. Not everybody thought the Afar and Laetoli hominids were the same thing; and some believed, with Richard Leakey, that more than one species was present in the Hadar sample. More personal complications also immediately arose. According to Johanson, Mary Leakey had agreed to join him, White, and Coppens in describing the new species, on the predictable condition that the announcement not suggest that any *Australopithecus* had been ancestral to the genus *Homo*. This, of course, would mean eliminating any offensive potential comparisons. But by the time the paper announcing the new name was in proof, Mary had nonetheless taken umbrage at the thought of "her" fossils from Laetoli being given the name of a different place, or perhaps not being placed in *Homo*. She demanded that her name be removed from the list of authors—something that could only be accomplished by destroying the entire first print run of the publication, and reprinting. Mary had her way, but only at the cost of near–total alienation from her would-be coauthors.

Unfortunate as it was, this development nevertheless had the advantage of giving Johanson and White a free hand to publish their conclusions, which duly appeared early in 1979 in the prestigious journal *Science*. After discussing various potential permutations, Johanson and White opted for a simple bifurcating scheme in which the stem species *Australopithecus afarensis* gave rise, at some point following 3 million years ago, to two lineages. One of them led, via *Homo habilis* and *H. erectus,* to *H. sapiens.* The other saw the gracile *A. africanus* dead-ending in the *A. robustus,* which Johanson and White did not distinguish from its East African variant, *A. boisei.* This scheme looked highly sophisticated at the time, and in its unrelenting minimalism it clearly betrayed Ernst Mayr's enduring influence on paleoanthropology. Yet for some, it wasn't minimalist enough. The distinguished South African paleoanthropologist Phillip Tobias, for one, claimed that East African *A. afarensis* was no more than a local variant of his own South African *A. africanus* (which was the actual progenitor of all later hominids). At the other end of the spectrum, though, Todd Olson of New York's City College concluded that most of the Hadar specimens, along with those from Laetoli, were robust australopiths, for which he preferred Robert Broom's genus name *Paranthropus.* Asserting that these fossils belonged to the species Kohl-Larson had discovered at Laetoli back in the 1930s, and that Weinert had named *Meganthropus africanus* in 1950, Olson claimed that they should take the name *Paranthropus africanus.* On the other hand, he claimed that Lucy and certain other very small Hadar individuals were distinctive. In his view they should be assigned to the genus *Homo,* which he extended to embrace the South African graciles. Under the rules of zoological nomenclature, which give priority to the name first published, all of the latter fossils together represented the species *H. africanus.*

Oddly, this radical divergence of opinion brought us back to the old dichotomy of *Homo* versus something else, variously termed *Australopithecus* or *Paranthropus.* This dichotomous mind-set is something from which paleoanthropology has proven very reluctant to free itself—even as the argument continues, albeit mostly at a muted level, about how many species of hominid are represented at Hadar. The majority of paleoanthropologists today still believe, with Johanson and White, that there is only one species in the combined sample, while the minority who consider that there is more than one disagree on exactly how the specimens should be allocated, and on what they should be called. Their predicament reminds us strongly of just how difficult basic systematics is. Sorting the specimens you have before you—or worse, those scattered around the museums of the world—is actually one of the most difficult tasks a paleobiologist of any kind ever takes on. Much too frequently, paleoanthropologists conveniently regard it as a simple filing operation that has to be got out of the way—or, far too often, brushed aside—before more important matters such as adaptation can be addressed. Yet as I've already observed, if you ever want to understand the evolutionary play you have to know who the actors are, as well as the parts they are playing.

BIPED, CLIMBER, OR BOTH?

Nonetheless, it was what the hominids were doing, and not who they were, that mainly preoccupied paleoanthropologists during the decades following the formal announcement of the bipedal *Australopithecus afarensis*. At that point, the assumption persisted that bipedality had evolved in the context of freeing the hands to make tools and to carry things around. Because of this, the discovery that Lucy and her kind well predated stone toolmaking clearly mandated some rude rethinking. Influential in this process was the Kent State anatomist Owen Lovejoy, who led the detailed description of the Hadar postcranial fossils. Particularly because of the exaggeratedly humanlike width of Lucy's pelvis, he rapidly became convinced that *A. afarensis* was a very efficiently adapted upright strider. Combined with features of the femur, Lucy's great pelvic breadth suggested to him that *A. afarensis* had possessed a very efficient muscular mechanism for stabilizing the hip in an upright position. What's more, Lovejoy came to believe that this apparent hyperadaptation to bipedality had been possible because the small-brained Lucy had not been subject to the same obstetrical problems that large-headed modern humans are (in many parts of the developing world the complications of childbirth are still a leading cause of death among young women). In Lovejoy's view, later humans had sacrificed efficiency in bipedal locomotion in order to have their big brains.

All very well; but why had Lucy and her kin become bipedal in the first place, if not to make and use tools? Rejecting the notion that the transition from quadrupedality to bipedality could have taken place in a single giant anatomical leap, Lovejoy sought a factor that might consistently have governed a long-term transition between the two locomotor styles. And in good neodarwinian fashion he concluded that everything had been directly related to reproductive success. The logic went like this. On their own, there was not much that *Australopithecus afarensis* females could do to improve their reproductive rates and give themselves evolutionary advantage. After all, they already had their hands full obtaining enough food for themselves and their encumbering offspring. But they *could* cope with more children if they contrived to co-opt a free-ranging male to help feed them. The suspicious male, however, would only bring home the bacon if he could be sure the offspring whose prospects his efforts enhanced were indeed his own. So, besides carrying food and ranging around home bases, early hominids must also have had a social system that encouraged permanent pair-bonding and sexual fidelity. This bonding would have required an elaborate system of sexual signaling. And in bipeds this would have involved the evolution of such otherwise inscrutable secondary sexual features as prominent breasts, pubic hair, and concealed ovulation.

Lovejoy's hugely complex narrative, articulated at some length in 1981, was predictably controversial. It did, though, have the salutary effect of reopening discussion of the putative advantages of bipedality, and the following decade or so saw several important studies of the comparative efficiencies of bipedality versus quadrupedality on the ground. The advantages of upright posture in maintaining constant body and (critically) brain temperatures in an open tropical environment were also usefully sketched out for the first time. Still, the very basis of Lovejoy's complex biosocial argument was rapidly undermined by further anatomical studies that questioned the degree to which *Australopithecus afarensis* was actually committed to the ground. Two French paleoanthropologists, Brigitte Senut and Christine Tardieu, concluded that the mobility of the upper and lower limb joints was greater in *A. afarensis* than in modern humans, suggesting enhanced climbing capacities—exactly as did the high mobility of the wrist joints that was noted by Henry McHenry of the University of California,

Davis. Bill Jungers of Stony Brook University observed that Lucy's relatively short legs would have favored climbing, while Yoel Rak of Tel-Aviv University subsequently noted that her widely flaring pelvis, instead of being a badge of superefficient bipedality, might have been a simple mechanical consequence of those short legs, compensating for them by exaggerated rotation. Russ Tuttle of the University of Chicago found that the hand and foot bones in the First Family collection were rather long and curved, indicating strong grasping capacity and by extension some degree of arboreal behavior. Moreover, it's probably safe to conclude, with Will Harcourt-Smith of the City University of New York, that the short and remarkably humanlike Laetoli footprints are very unlikely to have been made by the hominid whose foot bones were found at Hadar.

Putting all of the available evidence together, Jungers and his Stony Brook colleagues Randy Susman and Jack Stern proposed that, while *Australopithecus afarensis* was without question bipedal when moving around on the ground, it did not necessarily spend most of its time there. Capitalizing on their small body size, these early hominids almost certainly remained dependent on the trees for shelter, especially at night, and would presumably have spent a lot of time foraging there, too. Just what the exact balance between the ground and the trees might have been is anybody's guess. The derived structure of Lucy's pelvis and leg leaves no doubt that she spent a lot of her time as a biped on the ground. But at the same time there is no doubt that *A. afarensis* was descended from an at least primarily arboreal ancestor. There is no way to be sure to what extent the primitive arboreal attributes of the species were simply "baggage" inherited from that ancestor, as opposed to features that actively influenced its lifeway. But, since any organism has to survive as the product of its overall structure, it is a good bet that Lucy and her like spent significant amounts of time both on the ground and in the trees, with the result that these hominids are often seen merely as "habitual" bipeds. This contrasts with "obligate" bipeds such as we are (while chimpanzees, which can walk upright if they want to, but don't often do so, are "facultative" bipeds).

A "have your cake and eat it" lifestyle of the kind suggested by Jungers and colleagues is certainly implied by the environments in which *Australopithecus afarensis* normally lived. Hadar today is mostly a harsh, arid desert, but in Lucy's time it was a place of meandering rivers bordered by gallery forests and marshes that gave way to woodland, bush, thickets, and areas of open grassland. The dry, open ancient environment of Laetoli was likely an unusual setting for hominids to be in—and even then, as you may recall, the hominids who left the footprint trails there were apparently heading directly for the more densely vegetated Olduvai Basin. In a mixed habitat of the sort A. *afarensis* seems typically to have occupied, Lucy and her kin almost certainly exploited every resource open to them, including those available only in the forests, and uniquely in more open areas.

This opportunistic existence was evidently a very viable strategy for the early hominids, for as far as we can tell the resulting basic lifeway endured for several million years, even as various early hominid species came and went. And there is a lesson in this fact, because we are often tempted to see the australopiths as somehow transitional between completely arboreal and completely terrestrial lifestyles. But "transitional" implies a neither-fish-nor-fowl existence, glimpsed ephemerally on the way from one stable state to another; whereas, in reality, the australopith lifeway was evidently a stable and successful strategy, entirely on a par with what went before and what came after. The australopiths had their own way of carrying out business, just as we have ours, which emphasizes how distorting it is to see these precursors as junior-league versions of ourselves. Our manner of doing things almost certainly fails to provide an appropriate frame of reference for reconstructing early hominid lifeways.

CHAPTER 7

MEANWHILE, BACK AT THE MUSEUM . . .

FOR ME, COMING TO NEW YORK CITY'S AMERICAN MUSEUM OF Natural History in 1971 was like immigrating to a new world. Yale had shown me what very smart people could do while clinging to ways of thinking that were approaching the end of their useful life; at the American Museum, new ways of thinking were being born. The ichthyology curators Gareth Nelson and Donn Rosen were busy evangelizing for cladistics, while Niles Eldredge and Stephen Jay Gould, who had recently left for Harvard, were about to shake the comfortable assumptions of the New Evolutionary Synthesis (also born at and around the Museum) to their very foundations. In a real sense my education was just beginning, and it was my great good fortune that one of the first new colleagues I met after arriving in New York was Niles Eldredge.

Although by then a trilobite paleontologist and evolutionary theoretician, Eldredge had started out with an interest in anthropology, so maybe it is not surprising that soon we found ourselves collaborating on a study of the human fossil record. At the time, Milford Wolpoff was loudly proclaiming the single-species message in its purest form, and the fossil evidence that was eventually to refute this notion beyond any doubt was not yet at hand. Yet when Eldredge and I looked at the evidence available to us from the new perspective, it rapidly became clear to us that the signal emanating from the hominid record was the very reverse of the one predicted by the gradualist model.

In a book chapter published somewhat tardily in 1975, long after I had left for a year's lemur fieldwork in Madagascar and the Comoros, we pointed out that, since the fossils that are the archive of evolutionary histories must necessarily be physically discovered, it had been tacitly assumed that evolutionary histories themselves were matters of discovery. The unspoken feeling was that if you crawled over enough rocky outcrops and found enough fossils, the course of evolution would somehow be revealed to you. And of course, under the gradualist view that sees fossils as links in a chain running through time, this was a fairly plausible expectation. After all, it seems pretty self-evident that if you find enough of those links, and date them properly, you will effectively have the chain itself. It's hard to fault the logic. But logic never applies to the real world if the initial assumption is false. And as we saw earlier, some paleontologists, at least, had long known that the picture of evolutionary process that underpinned this viewpoint was, if not entirely wrong, certainly far from complete. As you'll recall, Darwin himself had been deeply unsettled by the lack of the expected intermediates in the fossil record, a condition he ascribed to the latter's inherent incompleteness. It had then taken well over a century before Eldredge and Gould explicitly proposed that the apparent gaps in the record might actually contain information. Still, if Eldredge and Gould were right, and species really are individuals with births, lifespans, and deaths, then your perspective on the fossil record has to change dramatically. The fossils that confront you actually sample a complex genealogy of species, each one produced at some point in the past by the splitting of a lineage. For every fossil species you discover, there is another one out there—whether known or as yet unknown from tangible fossils—to which it is most closely related by descent. The resulting pattern of relationships among your fossil species and their putative living descendants is not something that can be directly discovered: it has to emerge from analysis. More is required than mere discovery. So the question arises: What qualities of your fossils should you look at when you are trying to understand their places in the Tree of Life, and how should you analyze them?

From the point of view of the systematist, a fossil has basically three attributes: what it looks like, where it comes from, and how old it is: properties that are of unequal value when you are trying to reconstruct evolutionary histories. Where you find a fossil species may well be relevant to its adaptation, but it has no necessary bearing on what other species it is most closely related to. Similarly, you have to be cautious when weighing a fossil's age, for you can never be certain of the longevity of the species to which it belonged, or what part of its species' age range your particular fossil represents. And if you accordingly exclude age and geography as indicators of a fossil's identity and relationships, you are left only with morphology (nowadays interpreted in the broadest sense, to include molecules). This is the sole property of a fossil that bears an unequivocal imprint of its history. As we saw in chapter 5, it is not always easy to unravel that imprint in practice; but to the extent that we can reconstruct the hierarchy of primitive and derived character states that evolution produces with the data available to us, we can make testable statements about relationships within an ever-widening circle of groups. In a pinch, paleontologists may occasionally use the age of a fossil as a very general guide to the primitiveness or derivedness of its features, but such inferences remain untestable and are never better than vague statements of likelihood.

Beyond this, even the testable statements you *can* make are limited to the kind of relatedness that is due to descent from a common ancestor. They are simply about closeness of relationship, and not about ancestry itself. Indeed, if you wish to claim that a known species A gave rise to a later species B, then you would have to show that A conforms to the reconstructed morphology of the common ancestor of the pair in every detail, something highly unlikely to be the case. What is more, in the unlikely event that A *were* primitive relative to B in every feature, there would be no derived character to link the two and thereby demonstrate relationship. In other words, whenever you make statements about ancestry and descent, you are moving beyond the testable confines of the cladogram. No harm in doing that, of course, just so long as you realize that an imaginative leap is involved. And of course, if you can show that an earlier species possesses derived characters that are not present in a later one, you can confidently rule it out of ancestral status.

Having set the scene in this way, Eldredge and I proceeded to look at the human fossil record. At the time, neither of us had any close familiarity with the original fossils that document hominid evolution, but we both came to whatever evidence we had with a systematist's eye, gained from working with groups that were significantly diverse. And what we saw was certainly not continuity. Using a rather limited character set, we constructed a cladogram of hominid relationships—the first of many—that reinforced this initial perception. Those were early days, and our units of analysis reflected many holdovers from tradition: we included Neanderthals and modern humans, for example, as subspecies within a single species, as Dobzhansky and Mayr had advocated; and we still accepted *Ramapithecus* as an outlier within the hominid family. Conversions are rarely Damascene! But some of the conclusions that resulted from our preliminary exercise were eye-opening, at least to us.

Most interesting of all was the extreme difficulty we encountered in viewing *Homo erectus* as an ancestral morphotype for *H. sapiens*. In the structure of its cranial vault, *H. erectus* very evidently possessed a variety of derived characters that it did not share with *H. sapiens*. Indeed, we concluded that this species had assumed its traditional position as "hominid in the middle" solely because it was found in the "correct" stratigraphic position, intermediate in time between the gracile australopiths and *H. sapiens*. It hadn't hurt, either, that this far-flung eastern Asian form had been the first really ancient hominid species to be discovered, assuming its canonical position in the paleoanthropological pantheon very early on. We pointed out that, in terms of morphology, it was easier in many ways to derive *H. sapiens* from Louis Leakey's *H. habilis* than it was to view it as a descendant of *H. erectus*. This was true as far as it went (which in hindsight was not that far), but although I wish I could report that paleoanthropology took the implicit lesson immediately on board, our paper sank with few ripples. Still, we had made the point that only once a cladogram has been established for a given group of organisms can what we *know* about their relationships be distinguished from what we simply *believe*. And we did point to "a basic weakness in our present knowledge: we do not know precisely how many taxa we are dealing with in our discussions of the hominid fossil material, or what all their characteristics are."

That issue of how to recognize species in the fossil record became the focus of much of my later career. But in the interim, after I had returned from my long spell of lemur fieldwork with all its attendant mishaps, Eldredge and I followed up our initial, rather obscure book chapter with an article in the more widely read *American Scientist*. There, we more explicitly contrasted the simple basic cladogram with other kinds of evolutionary

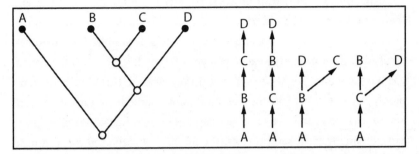

Diagram showing how cladograms and phylogenetic trees differ. As discussed at length in Chapter 5, cladograms such as the one on the left are simply statements of generalized relationship: B and C are more closely related to each other by descent than either is to D, while A is genealogically equidistant from all. The tree is a more complex statement that may also involve time, as well as ancestry and descent; and all the trees shown on the right are both compatible with, and can be derived from, the single cladogram at left.

hypothesis. We reiterated that a cladogram carried no implications beyond the nesting of organisms according to their character states and, by implication, their genealogical relationships. We also pointed out that the traditional "phylogenetic tree" represents a more elaborate level of analysis, because it specifies the nature of the relationships among the species of interest. In a cladogram the ancestral forms represented at the branching points are entirely hypothetical. They are characterized simply by a list of the character states expected at that point. All taxa included are "terminal," which is to say that all of them lie side by side at the top of the cladogram, with fossil and living taxa treated in exactly the same way.

In a tree, on the other hand, relationship is more nuanced. Two different kinds of relationship may be recognized in a tree: that between an ancestor and its descendant, and that between two "sisters" derived from the same ancestor as a result of lineage splitting. In practice, choosing between these two possibilities demands a judgment call; huge difficulties are usually attendant on recognizing both ancestors and speciation events, and it cannot be proven that two taxa are the immediate daughters of a given parent. And if you add the time dimension to the tree, as you inevitably do when you include fossils, then you find yourself with a very complex and not easily testable statement. Indeed, as the figure above shows, a single cladogram can potentially be transformed into a number of different trees depending on how you depict the relationships involved.

Nonetheless, while your phylogenetic tree is not testable, to the extent that it adds subjective judgment to the basic information contained in the original cladogram it is nonetheless a considerably simpler kind of statement than the scenario you get when you add in all the other information you have about the fossils in your tree. Information of this kind most commonly relates to what you know about the environment your fossil lived in, and to what you can infer about its adaptations. And just as one cladogram can potentially yield several trees, the number of scenarios you can base on a given tree is almost limitless. By the time you have formulated your scenario, you are so far away from anything directly testable that the main determinant of its plausibility will be your storytelling ability.

Of course, precisely because of their complex content, scenarios are incomparably the most interesting kind of evolutionary statement you can make. And it would be ridiculous to object to them just because they are not easily weighed and evaluated. So what Eldredge and I emphasized was that, just as any tree should be based on a cladogram, every scenario should be based on a clearly articulated tree. We pointed out that paleoanthropologists spent a lot of time talking straight past each other because they typically jumped in at the deep end, going directly to their preferred scenarios without specifying the simpler formulations on which they were based or, more likely, without completing the necessary preliminaries at all. Shouting matches—whether muted or shrill, whether on the front pages of newspapers or in small print in arcane trade journals—were the inevitable result.

SHOUTING MATCHES

The most famous paleoanthropological shouting match of all occurred on national television, between Richard Leakey and Donald Johanson. Throughout the latter part of the 1970s, relationships had steadily deteriorated between the pair, and by the end of the decade they were barely on speaking terms, although both benefited enormously from the publicity generated by their rivalry. Eventually their sniping became so notorious that, in 1981, the CBS television network organized a prime-time Leakey-Johanson debate, to be broadcast live from the American Museum of Natural History and moderated by no less than the heavyweight newsman Walter Cronkite. The first participant to arrive at the museum was Johanson, debonair and fearsomely prepared. Leakey was disgorged at the last minute from a limo, disheveled and exhausted after a long flight from Kenya and, by his account, unsure of what to expect.

Soon the participants were settled into chairs in front of an array of reconstructed heads of extinct hominids. The klieg lights were turned on and the fun began. Cronkite started with the expected round of congratulations on the fabulous fossils both his interviewees had found, and invited them to make the usual airy statements about the importance of the work they were doing. He then cut to the chase, and began inquiring about the scientific beliefs of his guests and the reasons for their vociferous disagreements. Thus began the high drama.

As Leakey held forth, fluently disguising his own lack of any coherent paleoanthropological framework with sly disparagement of his adversary, Johanson reached beneath his chair. With a grand theatrical flourish, he produced a large white board and a black felt marker. On one side of the board was a beautifully drawn version of the hominid family tree that he and White had published not long before. The other side was blank. Proclaiming that here was his own very specific statement about human evolution, Johanson thrust the chart and the marker into Leakey's hands, challenging him to draw his alternative version in the blank space. Justifiably nonplussed, after a moment's hesitation Leakey grabbed the chart and drew a large X across the Johanson-White tree. When Johanson asked him what he would replace it with, Leakey angrily responded, "A question mark!" Drawing a large interrogation symbol on his side of the chart, he handed it back to Johanson and stalked off camera, leaving Cronkite, me, and everyone else staring in astonishment. The rupture between the two stars of paleoanthropology was complete, and it lasted for 30 years until mutual advantage and the mellowing of age once again brought them together in a bland and painfully polite replay at the American Museum in 2011.

This incident was not exactly typical of the level of discourse in paleoanthropology at the beginning of the 1980s, but it did provide a sort of metaphor for how paleoanthropological business was conducted in those days. Basically, one expert's declaration was pitted against another's, and authority in such disputes was conferred by control of the new fossils that underwrote those authoritarian revelations. The costs of entry into the debate were high, and those whose voices were heard were those whose good fortune (and, it must be added, usually hard work and dedication as well) had brought important new fossils into their hands. Any argument was much more likely to be listened to if it was based on a new and preferably distinctive fossil; and keeping such fossils under tight control was a good way of ensuring such arguments would remain undisputed. If you were old-fashioned and studied, say, Neanderthals, a paleoanthropological legacy of which by then numerous examples were known, your new ideas might be listened to even if you had no new fossils to contribute to the pot. But in the fast-moving world of new discoveries in the Pliocene and early Pleistocene, you basically had to pay to play, in fossil currency.

Much of the mind-set generated at this crucial time in the history of paleoanthropology still lingers, although gradually the value of ideas, and attention to how they are formulated, is coming to be more widely appreciated among students of human evolution. For a whole number of reasons change is finally afoot, and at least the jargon of cladistics has found its way into paleoanthropology. But meanwhile, one big area of concern still remains basically unaddressed, and even largely unacknowledged. This is the crucial matter of how we should recognize and delimit species, the basic actors in the evolutionary play. In our 1977 paper Eldredge and I had already complained about the Synthesis-inspired "fashionable reluctance to apply species names to some of the more recently discovered hominid fossils," notably those then emerging from Kenya and Ethiopia. We noted that, typically, the differences visible in the bones and teeth of primate species in the same genus are rarely very striking, and we implied that an understanding of diversity in the hominid past would be well served by bearing this in mind. This remains as true today as it was then, but the human anatomist's austere, minimalist mind-set continues to dominate, and paleoanthropology still struggles with this most basic of issues.

OUTSIDE AFRICA

While Africa was grabbing public attention, important new discoveries of fossil hominids were being made elsewhere, with less fanfare. In Java, the first *Homo erectus* cranium with a face attached was found in 1969. Known as Sangiran 17, for the area in which it was found, it boasted a braincase with a volume of over 1,000 ml and had an unexpectedly massive face. Subsequently, a number of other *H. erectus* fossils were found in central Java, and efforts were begun to obtain radiometric dates. Such dating has proven tricky, but nowadays it is reckoned that the original Trinil skullcap is around 700,000 years old, while much of the material from Sangiran, including Ralph von Koenigswald's early finds, dates from about a million years ago. However, potassium-argon dates also indicate pretty firmly that *H. erectus* was around in Java as early as 1.6 million years ago, and even possibly a bit before that. Perhaps most remarkably of all, a series of skulls discovered before World War II at a place called Ngandong was eventually dated to an amazingly recent 40,000 years ago. The Ngandong hominids have rather larger braincases than those from Sangiran and Trinil (the largest of them is a whopping 1,251 ml), but they are clearly of the same general kind as their precursors, and are without doubt the most striking exemplar we have in paleoanthropology of Giambattista Brocchi's dictum that species sometimes have very long lifespans.

In China, excavations at Zhoukoudian during 1966 produced part of a *Homo erectus* skull, which actually articulated with a cast of a cranium

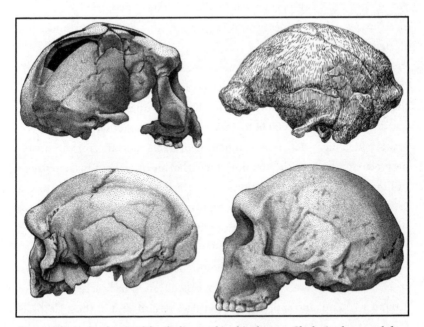

Some of the earlier hominid fossils discussed in this chapter. Clockwise from top left: the Sangiran 17 Homo erectus *cranium from Java; a late* Homo erectus *skullcap from Ngandong, Java; the* Homo heidelbergensis *cranium from Petralona, Greece; the* Homo heidelbergensis *cranium from Dali, in China. To scale. All drawn by Don McGranaghan except top right, by Diana Salles.*

that had been exhumed there before World War II, but had subsequently been lost with the rest of the collection. At around the same time, the application of radiometric dating methods began to shed light on how old Peking Man was. Early estimates ran from about 230,000 to 460,000 years, but subsequent determinations have tended to be a little older than this. The most recent work, principally using a method known as cosmogenic ^{26}Al/^{10}Be (aluminum/beryllium), suggests an antiquity of between about 680,000 and 780,000 years. As used at Zhoukoudian, cosmogenic ^{26}Al/^{10}Be works by determining the ratio in a sample of quartz grains of two different radioactive nuclides that decay at different rates. The ratio is fixed when the sample is exposed to sunlight, but when it is buried, as happens to sand grains that become incorporated into an archaeological deposit, the ratio starts to change as one nuclide decays faster than the other. Other things being equal, the change in ratio is proportional to time. Dates made earlier, during the 1980s and 1990s, had mainly been derived from then-new techniques such as electron spin resonance (ESR), and thermoluminescence, which depend on the fact that free electrons released by natural radiation get trapped in mineral crystals at predictable rates. If you can count those electrons, you have a measure of how much time has elapsed since they started accumulating.

Each technique has its own uncertainties, so to put the rather large date range from Zhoukoudian in perspective, a couple of crania very similar to the Zhoukoudian ones, but found a few hundred miles away in Nanjing, are dated by ESR to about 400,000 years ago, although another technique (U-series, which measures the accumulation of stable thorium from an unstable precursor in the freshwater limestones that are often found as flowstones in limestone cave deposits) makes them considerably older. In contrast, a crushed skull found to the south at a place called Lantian may date from well over a million years ago. So despite a good measure of uncertainty, *Homo erectus* in all probability had a long run in China as well as in Java.

Still, not all middle Pleistocene hominids from China are *Homo erectus*. In 1978 a cranium was found at a place called Dali. It has a largish brain of around 1,200 ml, and sports tall browridges reminiscent of those decorating the African Kabwe cranium. Originally described as an *H. erectus* by its Chinese discoverers, it was later given its own species name, *H. daliensis*. It is also possible that the same sort of browridge was also originally present in at least one of a couple of crania from Yunxian, in Hubei Province, that are

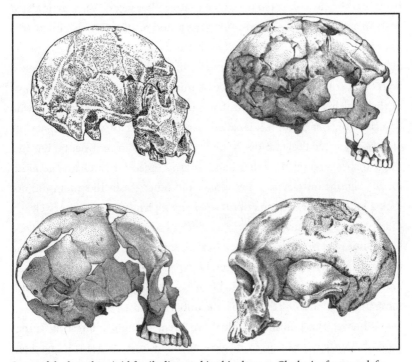

Some of the later hominid fossils discussed in this chapter. Clockwise from top left: Skull 5 from the Sima de los Huesos at Atapuerca, Spain; adult Neanderthal cranium from Amud, Israel; Jebel Irhoud 1 cranium from Morocco; Jebel Qafzeh 9 cranium from Israel. To scale. All drawn by Don McGranaghan, except top left, by Jennifer Steffey.

dated to around 400,000 years ago. In contrast, a skullcap found in the late 1950s at a site called Maba has rather delicately arching browridges that curiously recall those of the European Neanderthals. The status of this fossil is still debated. Nonetheless, however all of these fossils may ultimately be classified, twin signals of both diversity and longevity were already emerging by 1980 among Chinese hominid species of the middle Pleistocene.

Something similar was also happening in Europe. In 1960 a well-preserved cranium was found in a cave at Petralona, in northern Greece. Just like the Dali skull, its tall and backwardly twisting supraorbital ridges and vault volume of around 1,200 ml reminded many of the Kabwe cranium from Zambia. Sadly, its date remains very imprecise, though it may be in the range of 150,000 to 250,000 years. A lot older at around 450,000 years, but morphologically quite similar, is a hominid facial skeleton discovered in 1971 at the cave of Arago in France's eastern Pyrenees. When articulated with an apparently associated vault bone found nearby, this specimen produced a brain size estimate of between 1,100 and 1,200 ml. The Arago site also delivered some mandibles found within a deep stack of archaeological layers that contained an abundance of animal bones and stone tools, most of them fairly crude flakes, with few handaxes. The French scientists who described the Arago hominids called them *Homo erectus,* which at that time was the preferred designation for fossils of this kind in continental Europe. English speakers, on the other hand, mostly opted for "archaic *Homo sapiens.*" Significantly, Arago, Petralona, and other fossils like them did not look a lot like either *Homo erectus* or *Homo sapiens.* But this nicety clearly did not count for much in a world in which expectations were still governed by the linearity of the Synthesis. After all, something intermediate in time between those two species could not be expected to look exactly like either, and the name you chose depended largely on which species you thought it more unlike. Under the Synthesis, the important thing was the lineage.

Whoever the hominid might have been, though, interesting things were obviously happening in its general time range. Before starting work at Arago, the French archaeologist Henry de Lumley excavated a hominid occupation site on an ancient beach at Terra Amata, in the suburbs of Nice. His team uncovered the remains of what had evidently been a seasonal hunting camp whose occupants had constructed quite elaborate shelters some 380,000 years ago. The best-preserved of these shelters was an oval hut-like structure made of saplings that had been cut, stuck into the ground, and brought together at the top. The periphery was reinforced by a ring of stones, a break in which at one end indicated where the entrance had been. Just inside that entrance was a shallow, scooped-out depression containing burned stones and animal bones. At the time this was the earliest example known of an ancient fireplace, and it furnished the earliest really good evidence that hominids were controlling fire. A couple of much earlier sites in Africa had produced suggestions of shelter construction and fire use, but both were hotly contested, and it is not until recently that there has been good evidence of fire domestication in Africa at a million years ago. What does seem significant is that Terra Amata is situated at just about the point when fire began to be a regular feature in the archaeological record; it is thus emblematic of the emergence of human dependence on fire as a routine part of the hominid tool kit. We will see later that there are good circumstantial reasons for believing that fire use may in fact have played an essential role in hominid life much earlier than this, but as far as tangible evidence is concerned, it is Terra Amata that lies at or near the inflection point.

Still, even back in the early 1970s, sites like Arago and Terra Amata were not the earliest signs of hominid presence in Europe. Otto Schoetensack's Mauer jaw was long believed to date back to over half a million years ago (and a recent absolute dating, not accepted by everybody, has pushed it back to over 600,000 years). Earlier still was a handful of archaeological sites containing very simple stone tools, though even they eventually turned out not to document the very earliest hominid presence in Europe. We will get to that later, but at the time it appeared that hominids had finally emerged from their birth continent of Africa only at quite a late date. The earliest firm evidence for hominids outside Africa then came from an Israeli site called 'Ubediya, where handaxes believed to date from a little over a million years ago had been found. This great age appeared to be a bit of an anomaly since handaxes had made a pretty late appearance in Europe (and were totally unknown at the time in eastern Asia); but it was nevertheless pretty good proof that by this point handaxe makers were ranging beyond Africa, where the new technology had appeared well over half a million years earlier.

Almost as imperfectly understood was just how, where, and when the practice of making utensils by flaking a piece of rock to a particular form yielded to a subtler approach. But it was known that, at some poorly dated point, stone toolmakers had begun carefully shaping a core of predictably fracturing stone until a final blow would detach a blank that could easily and quickly be modified into a specialized tool: maybe a point or a scraper, or even a slender handaxe. A huge advantage of this technique was that the flake would have a continuous cutting edge around its periphery. As always, older-style tools persisted (especially where high-quality stone was not available) alongside the new ones, which began to appear at some time around 300,000 years ago. The new approach meant that good raw materials assumed a greater importance than ever, and toolmakers began to conserve them jealously, resharpening them and changing their shape, size, and apparent function in the process. Back in the 1970s, not many archaeological sites were known that represented early industries of this kind, which became collectively known as Middle Paleolithic in Europe, and as Middle Stone Age in Africa. But what was already evident was that the transition from handaxe-type industries to those based on "prepared cores" had not been smooth and continuous, and that the appearance of the new way of doing things had not been obviously linked to the arrival of any specifiable new kind of hominid.

Also in doubt was just when the Middle Paleolithic ended. This uncertainty was largely due to the intrusion into the western European archaeological record, between about 44,000 and 40,000 years ago, of the Châtelperronian industry. Interpolated between the Mousterian of the Neanderthals and the Aurignacian of the first modern humans to enter Europe, the Châtelperronian, found at a handful of sites in France and Spain, incorporated aspects of both traditions. In particular, the Châtelperronian stone tool kit contained a high percentage of the long, thin "blades" that were characteristic of Cro-Magnon industries; and at the French site of the Grotte du Renne at Arcy-sur-Cure, some clearly decorative objects of bone and antler were thought to be Châtelperronian—though there are now severe doubts about that particular association. For a long while, the biological identity of the makers of the Châtelperronian remained obscure, but in 1979 a fossil found at a French Châtelperronian site called Saint-Césaire tipped the balance in favor of the Neanderthals, and the issue was finally settled by the identification of Neanderthal fossil fragments at Arcy as well. The Châtelperronian was evidently the last gasp of the Neanderthals, rather than a harbinger of the Upper Paleolithic. Or not quite the last gasp: at a couple of sites, full-blown Mousterian overlay Châtelperronian levels. So there is evidently something quite complex to explain here. In the light of very recent molecular findings, it is just possible that the Châtelperronian was the result of a brief episode of Neanderthal acculturation by Cro-Magnons; but if so, the exact circumstances remain entirely mysterious.

In any event, the ultimate practitioners of prepared-core toolmaking were the Neanderthals, whose record significantly expanded during the second half of the twentieth century. In 1976 the back of a skull that had clearly belonged to a Neanderthal was found in a Mousterian context at Biache-Saint-Vaast in northern France. Dating from around 175,000 years ago, this was the oldest clearly Neanderthal fossil known, although the German fragments described by Franz Weidenreich before World War II may extend the Neanderthal record to over 200,000 years. But that was simply the species itself. The larger group to which *Homo neanderthalensis* belonged had actually been around in Europe far longer than this. In the early 1990s a team of

Spanish researchers described some crania from a site known as the Sima de los Huesos (Pit of the Bones) at Atapuerca in northern Spain. These fossils have many Neanderthal features, without possessing all of them; they clearly represent a Neanderthal precursor. For a long time this earlier species was erroneously attributed to Schoetensack's *H. heidelbergensis*; and since the withdrawal of that attribution it has been floating in taxonomic limbo.

Two decades of excavation at the Sima de los Huesos, which bizarrely lies at the bottom of a vertical shaft deep within a large limestone cave, have yielded the fragmentary remains of some 28 individuals, of both sexes and all ages: the largest sample of an extinct hominid species than ever found in one place. How the bones accumulated is still the subject of debate, although the scientists who discovered them believe the bodies were thrown down the shaft by their companions, post mortem. The hominid remains in the Sima are jumbled up with each other and those of numerous other animals, notably cave bears, and only a single artifact has been found: a beautiful quartz handaxe, finely shaped and completely unworn, that its finders interpret—dubiously, in my view—as a ceremonial object with the same symbolic weight to its owners as it might have to a modern person. The age of the bones in the Sima is also contested, although they are probably around 430,000 years old.

Farther east, in 1961 Japanese researchers found a reasonably complete skeleton of a young male Neanderthal—with the largest fossil human brain volume ever found, at 1,740 ml—in the cave of Amud in Israel. Thermoluminescence (TL) dates indicate an age in excess of 50,000 years, comparable with another Neanderthal skeleton excavated nearby at Kebara in 1983. The 1960s also saw the discovery, at Jebel Irhoud in Morocco, of two crania and a juvenile lower jaw, now thought to be over 160,000 years old. Both crania have volumes in the 1,300 to 1,400 ml range, right around the modern average. Found in putative association with Mousterian stone tools, they were initially identified, wrongly, as African Neanderthals. Although clearly distinctive, the Jebel Irhoud hominids have more recently been compared to specimens from Israel, including the Skhūl fossils, and to those from a site near Nazareth called Jebel Qafzeh. First dug in the 1930s, then again in the 1960s and 1970s, Qafzeh produced several burials of Mousterian context containing skeletons of various ages and completeness. Two of these, including the adult Qafzeh 9, are clearly *Homo sapiens*, while the others are not. The skull known as Qafzeh 6, for example, has a large brain volume of 1,658 ml, but the anatomical relationship of its face to its braincase is highly atypical for *H. sapiens*. Still, most paleoanthropologists have been content to assign all the Qafzeh hominids to *H. sapiens,* while muttering about how "archaic" some of them are.

Concurrently, other parts of Africa were also producing important new finds. For example, the caves of Klasies River Mouth at the continent's southern tip yielded fragmentary but modern-looking Middle Stone Age fossils. Some of these may be as much as 120,000 years old and apparently represent the leftovers of cannibal feasts. They seem to be approximately contemporary with a much less modern-looking facial skeleton discovered at Florisbad, deep in South Africa's interior, during the 1930s. Border Cave, on South Africa's frontier with Swaziland, produced a *Homo sapiens* cranium that may possibly be 90,000 years old, while in 1973 archaeologists digging at Lake Ndutu in Tanzania found a distinctive partial cranium that, at between 200,000 and 400,000 years old, is probably broadly contemporaneous with both the Kabwe skull and a braincase found 20 years earlier at Saldanha on South Africa's southwest coast. The Kabwe and Saldanha specimens resemble each other reasonably closely, as both do Jon Kalb's Bodo skull; but the Ndutu skull seems to belong in another camp. These and many other findings make it clear that much of great interest in human evolution was being discovered throughout the Old World in the later Pleistocene as the twentieth century wound down, disguised as it might have been by the tendency to squeeze everything under the umbrella of *H. sapiens*. But, once more, the limelight was more or less monopolized by developments in Kenya and Ethiopia.

CHAPTER 8

TURKANA, THE AFAR, AND DMANISI

DURING THE EARLY 1980s THE ATTENTION OF RICHARD LEAKEY'S Turkana team began to drift toward the western side of the lake, where equally important fossiliferous exposures lay. At a site called Nariokotome, the amazingly complete skeleton of a young male hominid who had died in a swamp, face down, some 1.6 million years ago, was discovered in 1984. Somehow, his remains had escaped dismemberment by scavengers for long enough to become protected by accumulating muds and eventually fossilized. The Nariokotome Boy was not yet adult, showing the degree of skeletal maturation you might expect in a modern 13-year-old. But he had developed rapidly, and was quite mature enough to give a good idea of how adult members of his species were built. And in no way was he an intermediate form, as paleoanthropologists still in thrall to Mayr might have expected of an individual of his time period. In contrast to short and broad-in-the-beam "bipedal apes" such as Lucy, the boy was tall and slender, with long legs and short arms. His face and teeth were much reduced, and even compared to his bigger body his brain was significantly larger than that of any australopith, or even of a *Homo habilis*. In short, he was a striding biped who already bore the major skeletal hallmarks of modern people. Unlike his immediate predecessors, who were "habitually" bipedal, he was an obligate biped, shorn of most reminders of an arboreal past. About five feet three inches tall when he died, researchers first thought that the Boy would have topped six feet had he survived to maturity. Now, though, it seems he had probably completed most of his growth and wouldn't quite have hit that mark.

And, he wasn't modern all the way. Among other things, careful analysis of the growth increments in his teeth, under very high magnification, showed that despite his advanced stage of skeletal maturation the Nariokotome Boy had actually died at the calendar age of eight years, having developed fast compared to a modern child—although significantly more slowly than an ape. The prolonged developmental schedule of humans has the downside of leaving the young totally dependent on their encumbered parents for a long time. But the countervailing advantage is a long period during which children can absorb complex behavior patterns and technologies. In tandem with this, human children are born with brains that are relatively small compared to those of adults, but that grow much more after birth than those of apes do. It is now known that australopiths typically developed on a rather apelike schedule; and though the Boy was somewhere in between apes and modern people on this measure, his 900 ml brain probably did not have much expansion left to achieve.

Unsurprisingly, this skeleton shows other features of bodily structure that are not entirely modern. It is possible, for example, that the individual was actually a little less slender than the original reconstruction of his rather crushed pelvis suggests. Indeed, the very svelte pelvis of modern humans seems to have been a rather late acquisition, associated with our remarkably light body build. Earlier members of the genus *Homo* tended, in contrast, to have quite robust and somewhat flaring hips, something that is certainly true of a million-year-old pelvis reported some years ago from a site called Gona in Ethiopia. It is hard to imagine that the Nariokotome lad would have grown up to be vastly different from the Ethiopian form.

But irrespective of the details, the most striking thing about this adolescent is that he clearly belonged to a species that had abandoned the trees. For the first time, we see a hominid absolutely committed to the ground, with little left in his arm or shoulder anatomy to suggest much climbing in his repertoire. Still, he did retain some signs of the past, both in his rather conical thorax and in a rather upwardly oriented shoulder joint that might have limited his capacity to throw objects. If so, this was a significant

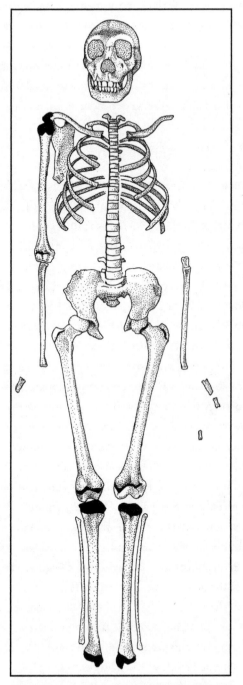

The adolescent skeleton KNM-WT 15000 from Nariokotome, West Turkana, Kenya. Drawn by Don McGranaghan.

limitation in his new terrestrial environment, for the savannas and bush-lands that were apparently his home teemed with predators, against which hurling rocks would have provided an obvious form of protection. None-theless, the Nariokotome Boy's radically altered anatomy is clear testimony to a new way of life away from the shelter of the trees. Almost certainly, this new lifestyle involved an increasing dependence on hunting.

BRAINS AND FIRE

While we can deduce for several reasons that by this stage hominids already relied significantly on the consumption of animal carcasses, a major one is that hominid brain expansion, which was really beginning in earnest dur-ing the Nariokotome Boy's period, comes at a considerable energetic price. Consider this: your modern human brain accounts for only about 2 percent of your body weight, but it consumes about one-fifth of all the energy you take in. Such low mileage has to be paid for, and animal fats and proteins are the most obvious currency in which to pay. Switching to that currency wasn't easy, however, for the environmental shift that must have been in-volved implies profound physiological changes. Life out in the open, ex-posed to the tropical sun, was almost certainly associated with the loss of the thick coat of body hair that apes still retain, and the beginning of body heat regulation by evaporative sweating. As long as they could maintain a supply of water, sweating allowed hominids to continue moving almost indefinitely while out in the full tropical sun—a trick that fleeter potential prey simply could not pull off. Eventually, this ability allowed our precur-sors to develop an entirely novel form of hunting, in which prey were relent-lessly pursued at slow speeds out in the tropic sun, and worn down as their heat loads accumulated. For no matter how fast it may be at normal operat-ing temperature, once an animal is overheated it begins to lose the ability to run away. One of the most significant trade-offs involved in adopting homi-nid bipedality in the first place had been the sacrifice of speed; but the new strategy placed not speed, but stamina, at a premium. Indeed, fast running on two feet over uneven and predator-infested country would have entailed high risks: one sprained ankle would have been enough to instantly convert a determined hunter into helpless prey.

Still, by heritage hominids are vegetarians. And turning yourself into a secondary carnivore is not easy. For example, while our chimpanzee cousins do hunt the occasional colobus monkey or juvenile bushpig, their motives for doing so seem typically to be social rather than dietary. One important reason for this may be that apes do not digest their meat intake particularly well. Indeed, for a creature—like us—that lacks a carnivore's specialized digestive tract, there appears to be only one really efficient way to increase energy extraction from animal fats and proteins: namely, to cook them before ingestion. Cooking almost anything makes contained nutrients more readily available to the digestive tract; and it also has the advantage of killing the toxins that accumulate rapidly in decaying carcasses under the tropical sun, rendering even scavenging a more attractive proposition. So great are the advantages of cooking, indeed, that the primatologist Richard Wrangham argued in his recent book *Catching Fire* that the ability to control fire, and thus to cook food, was essential for early *Homo* to invade the open savanna. Once there, fire would also have provided a degree of protection from all those open-country predators, especially at night (though it might have drawn unwanted attention, as well).

But while the circumstantial argument for early fire use is a strong one it is, alas, currently unsupported by empirical evidence. The first evidence of controlled fire comes from a South African site that is only about a million years old, while hearths only became a routine feature of hominid habitation sites about 400,000 years ago. Maybe this is because the traces of fire do not preserve well, in which case future technological refinements that allow us to spot them more readily may help clarify the question.

Despite their radically new anatomical structure and the changed lifeway it implies, the Nariokotome Boy and hominids like it were not accompanied by any material evidence of behavioral change. The tools eroding out of the sediments that yielded the earliest of them were basically indistinguishable from those associated with *Homo habilis* at Olduvai (which, based on a fragmentary skeleton found in 1986, had turned out to have a very archaic body structure) and with australopiths in Ethiopia. This lack of association heralded a pattern that was to become pervasive in Paleolithic archaeology: namely, that the arrival of new kinds of hominids rarely seems to coincide with the invention of new kinds of stone tool. Indeed, technological change itself was a rare event; when faced with environmental challenges, hominids apparently until very recently preferred to adapt old tools to new uses rather than to invent new tool types. Even with one isolated outlier at 1.78 million years ago, handaxes do not appear in any

meaningful numbers until a hundred thousand years after the Nariokotome Boy so poignantly fell into that swamp.

WHO WAS THE NARIOKOTOME BOY?

As sophisticated disciples of the New Evolutionary Synthesis, its excavators
attributed the Nariokotome skeleton to "early African *Homo erectus.*" This
followed the precedent already set a decade earlier for the 3733 and 3883
crania from the other side of Lake Turkana. It did not seem to matter very
much that the boy's skull did not look anything like the type materials of
Homo erectus from Java, or even that it was hard to imagine the individual
growing up to resemble either 3733 or 3883 very closely. After all, as everyone knew, individuals vary within the same species. And any species
that could contain such disparate individuals as the Nariokotome Boy and
the million-year-younger and geographically hugely distant hominid from
Trinil only seemed to prove that the sky was the limit. The Turkana researchers could certainly hope for history to be on their side, for the greater
the spectrum of materials already shoehorned within a particular species,
the easier it becomes to fit others in as well. Just before writing this, your
author encountered a pair of distinguished colleagues at a meeting, one of
whom had recently reported finding a huge trove of unidentified and only
very approximately dated hominids in southern Africa. The first question
the other asked him was, "Is it *Homo erectus?*"

So the trend continues, incessantly reinforced. In 2004 a group led by
the Smithsonian's Rick Potts discovered a partial cranium at Olorgesailie, a
Kenyan site famed for its abundance of handaxes. Slightly under a million
years old, the fossil is small and lightly built, and it is very distinctly different from both a penecontemporaneous cranium discovered at Buia, in
Eritrea (by Italian researchers who suggested it was somehow intermediate
between *Homo erectus* and *H. sapiens*), and from the Sangiran materials
from Java, also of about the same age. Neither does it particularly resemble
older Kenyan fossils such as the 3733 and 3883 crania. Yet its finders nonetheless attributed the Olorgesailie fossil to *H. erectus.*

An even more revealing mind-set was exhibited not long after this by a
group of researchers who, in 2007, finally reported evidence for two coexisting lineages of *Homo* at East Turkana. Discovered not far from where the
type mandible of *Homo ergaster* had been found, and similarly around 1.5

million years old, this evidence consisted in part of a maxilla, with greatly worn teeth, that the researchers attributed to *H. habilis*—which by that point had become just as much of a taxonomic wastebasket as *H. erectus.* The other lineage was represented by a very small cranium that the group thought was *H. erectus*—even though it boasted none of the major features that make the directly comparable Javan type specimen of this species so distinctive. This latter identification was particularly ironic, for while the researchers clearly took pride in having spotted this evidence for diversity (two *Homo* lineages coexisting at East Turkana!), the allocation of the cranium to *H. erectus* could only have been rationalized by viewing *H. erectus* as a comprehensive, Old World–wide, and potentially limitlessly variable middle stage in human evolution—precisely the construct they were purporting to disprove!

The year after the Nariokotome Boy was found, another startling hominid was picked up in older sediments a little way to the south. Known as the Black Skull from its dark patina, this 2.5-million-year-old specimen consisted of most of a robust australopith cranium without its teeth. Apart from being at least half a million years older than most other robusts, which were noted for the flatness of their faces, this one had a rather protruding muzzle, though it may originally have been a bit shorter than the current reconstruction suggests. Despite its manifest peculiarities, the new cranium was at first classified in *Australopithecus boisei,* but soon most agreed that it should be shifted to a species of its own. What that species should be called is a bit of a puzzle, though the tacit agreement currently is to link the Black Skull—rather improbably—with the penecontemporaneous small and similarly toothless jaw from Omo that Camille Arambourg and Yves Coppens had described in 1968. That would make the West Turkana specimen *A.* (or *Paranthropus*) *aethiopicus.* Despite the uncertainties, these specimens and some other fragments from Omo seem to demonstrate the independent existence of the robust clade back to at least 2.8 million years ago.

SEEN FROM AFAR

While all this was going on in Kenya, momentous developments were also taking place in Ethiopia, on both the political and the paleontological fronts. The years following the overthrow of the emperor Haile Selassie in 1974 were a time of enormous political violence and upheaval, leading ultimately

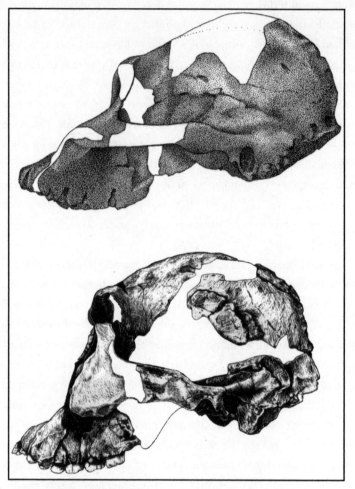

Above: The KNM-WT "Black Skull," assigned to Paranthropus
aethiopicus, *from Lomekwi, West Turkana, Kenya. Below: AL 444–1*
from Hadar, Ethiopia, the best-preserved cranium of Australopithecus
afarensis. *To scale. Above: Drawn by Don McGranaghan; below:*
adapted with permission from Kimbel et al. (2004; art by Yehudit
Sherman).

to the assumption of power by an unstable and rabidly anti-American military dictator. Less overtly, a bitter struggle broke out among paleoanthropologists for access to the Afar hominid and archaeological sites. This epic
saga of intrigue lasted for years, and involved a lot of unseemly behavior on
all sides. I vividly remember being loudly and publicly yelled at in 1995 for
having had the temerity to suggest that an academic session at a professional

meeting was an inappropriate place to make political accusations of claim jumping.

Thus it was that in late 1978, just after finding some more bits of the Bodo skull, Jon Kalb found himself given a week to leave Ethiopia permanently. Barely had he left, than the veteran Berkeley archaeologist Desmond Clark made a first foray to the Middle Awash. Soon after this, Donald Johanson and his colleagues found themselves banned from Hadar, while Clark and Tim White, by then also at Berkeley, formally acquired the Middle Awash concession. In 1981 the duo made the first in what would eventually prove to be a long string of key finds that spanned the last 6 million years. But the next year, in the midst of more political and other crises in Addis Ababa, all fieldwork in Ethiopia was put on hold.

When the moratorium on field prehistory in Ethiopia was finally lifted in 1989, both the Berkeley group and Johanson's Institute of Human Origins (IHO) resumed work in their respective field areas. But they did so in an atmosphere of mutual hostility that broke surface most notably in 1994, when the Berkeley group lured away from the IHO its major funder, its dating laboratory, and its icon, Clark Howell. This usurpation forced a major crisis that eventually led to the relocation of the IHO from Berkeley to Arizona State University in Tempe, and was exacerbated in the shorter term by those allegations of claim jumping I've already mentioned. These were levied against the IHO in 1995 by an Ethiopian student speaking for the Berkeley team. The incident was particularly sensitive because the site concerned, in the Gona River drainage to the west of Hadar, lay in the uneasy no-man's-land north and east of the classic Middle Awash sites. It had first been located back in 1973 by Gudrun Corvinus, a member of the original Afar survey group, and it contained the earliest stone tools known, between 2.5 and 2.6 million years old.

More lurid detail is hardly necessary to suggest that this was not paleoanthropology's finest hour. But despite all the tensions and bitterness, fabulous discoveries were made both at Hadar and in the Middle Awash during the 1990s and beyond. Following its return to Hadar in 1990, the IHO group, now led by the paleoanthropologist Bill Kimbel, discovered several dozen more *Australopithecus afarensis* fossils, most of them a little later in time than those that had been found back in the 1970s. Pride of place was taken by two fragmentary but reconstructible 3-million-year-old skulls, one belonging to a big male and the other to a much smaller female.

At 550 ml the brain volume of the male (known as AL 444–1) is greater than estimated for other *A. afarensis* (though it still lies within the general australopith range); but then, the individual himself was remarkably large, leading to the suggestion that members of his species may have become bigger over time. The face of this specimen is fairly robust and projecting, but the teeth are hardly massive in proportion to the size of the skull, and they retain the basic proportions seen earlier in the record.

Equally interesting was a find made right at the top of the geological section exposed at Hadar. The hominid palate concerned, dated to 2.3 million years ago, was assigned to a very early member of our genus *Homo*. This identification was facilitated by the presence in the same sediments of Oldowan-type stone tools, but the specimen nonetheless has pretty big chewing teeth, and might also be argued to have greater affinities with the *Australopithecus afarensis* fossils lying beneath it than with anything meaningfully attributable to *Homo*.

Meanwhile the Middle Awash badlands, while not as rich as those at Hadar, were producing fossils that covered a much wider time band. A thick stack of occasionally exposed sediments eventually produced bits and pieces of hominids ranging in age from 0.16 to 5.8 million years. Jon Kalb's expedition had already found the Bodo skull (then guessed to be about 350,000 years old) back in 1976, and on its first foray in 1981 the Berkeley group found a fragment of a second hominid skull, close to the original find site. Clark's team additionally recovered a hominid femur at a 3.4-million-year-old locality called Maka, a little way to the south, and some slightly older cranial fragments nearby at Belohdelie. Work was then suspended for nine years as a result of the general revocation of field permits, but when it resumed in 1990, a return visit to Maka also produced a well-preserved mandible and some other hominid bits, all of them attributed to *Australopithecus afarensis*.

By 1994, Clark and colleagues were also able to report a surprisingly ancient date of around 600,000 years for the Bodo skull, and showed that the sediments from which it came also contained Oldowan sites, the youngest known for this industry. However, handaxes appeared in the sediments immediately above the hominid level, prompting the speculation that the apparent Oldowan of Bodo was simply a variant of the Acheulean that happened to lack handaxes; this speculation reflected the difficulties inherent in classifying technologies in a world where the old routinely persists alongside the new, often for extended periods.

When they described it in 1978, the Kalb group had found that the closest resemblances of the Bodo cranium were with its counterpart from Kabwe in Zambia, and with the cranium found in 1959 at the poorly dated Petralona site in Greece. But rather than place these specimens together in a formal grouping, these researchers were content to consider that the Ethiopian fossil fell somehow into the "*Homo erectus–Homo sapiens* transition" that everyone in those days took for granted. Under the leadership of Tim White, who had studied under Milford Wolpoff and who became—for Middle Awash purposes, at least—an even more enthusiastic unilinealist than his adviser, the paleoanthropologists who took over the Middle Awash sites were only too happy to concur with their predecessors. Indeed, every specimen subsequently recovered from the Middle Awash somehow assumed its place in a steady gradation leading from the most primitive of hominids to the earliest *Homo sapiens*.

The next Middle Awash discoveries came from much lower in the geological section than the Bodo site. In 1992, some pieces of jaw were recovered from a 4.4-million-year-old site called Aramis. Containing suspiciously narrow molar teeth, these were given the name *Australopithecus ramidus* (later changed to *Ardipithecus ramidus*). Later on, a fractured and fragile partial skeleton also attributed to this form was found not far away, along with bits of several other individuals. Affectionately dubbed "Ardi," the skeleton (see reconstruction on the next page) was thought to have belonged to a female who had lived in a fairly densely wooded environment. At an estimated 110 pounds, Ardi weighed more than Lucy did, and she was a very different creature, as well as over a million years older. As reconstructed, Ardi's braincase is small, at under 350 ml in volume. But her canine teeth are reduced in the hominid manner, and some forward shifting of the foramen magnum testifies to upright posture, as do some controversially reconstructed features of her pelvis. Yet Ardi's flat feet had divergent and grasping big toes; and while the Middle Awash researchers concluded that she spent her time both in the trees and on the ground, they also contended that she had no suspensory adaptations in her skeleton—which is rather odd for a creature that was clearly too heavy just to have scampered along the tops of branches. Altogether, a bit of an enigma.

In 2001 and 2004 the time range of *Ardipithecus* was extended by the attribution to the genus of some bits and pieces that were found at various localities between 5.2 and 5.8 million years old. There was no way to

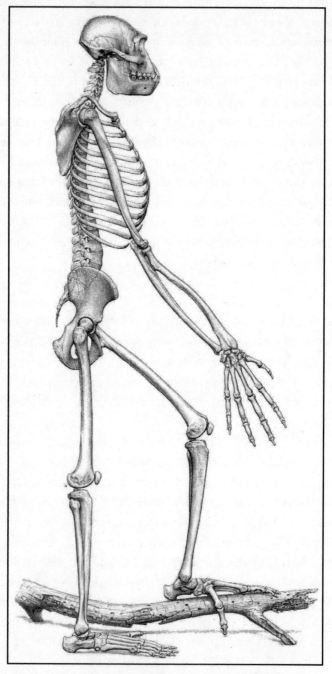

Reconstruction by the paleoartist Jay Matternes of the highly fragmented ARA-VP-6/500 Ardipithecus ramidus *skeleton from Aramis, Ethiopia. The most complete skeleton assigned to a very early hominid, this specimen shows a variety of unusual features, such as a divergent great toe. Courtesy of and © Jay Matternes.*

demonstrate with any certainty that all had in fact belonged to members of the same hominid species; but they were initially allocated to a new subspecies of *Ar. ramidus*, before being elevated to the separate species *Ar. kadabba* in 2004. Most of the fossils were isolated teeth—among them a couple of quite pointy canines—but the youngest of them was a toe bone that was claimed to show both apelike and humanlike characteristics.

KENYA AND CHAD

The obvious comparison for these materials was with the other contenders for the title of "earliest hominid" that had by then appeared on the scene. One of these was *Orrorin tugenensis*, named in 2001 by Martin Pickford and Brigitte Senut of the Collège de France from some pieces of jaw and femur shafts found the year before in northern Kenya. Dated to about 6 million years ago, the bluntish canine and squarish molars of *Orrorin* were plausibly those of an ancient hominid, and the partial femora were argued to have been those of bipeds. When they described their new finds the French researchers sidelined *Ardipithecus* to a hominoid branch ultimately leading to the chimpanzee, and also suggested that the hominid lineage itself had split, early on, to lead to *Orrorin* and ultimately *Homo* on the one hand, and to dead-end in *Australopithecus* on the other. Naturally enough this did not go down well with the Middle Awash group, who even questioned whether the Kenyan hominid was distinct from their Ethiopian candidate.

The other "earliest hominid" contender was *Sahelanthropus tchadensis*, represented by a cranium and some mandibular fragments found by another French team in the central African country of Chad in 2002 and 2005. These fossils were estimated to be between 6 and 7 million years old, and may be at the more ancient end of that range. What also made this discovery surprising was its location: some 1,500 miles west of the East African Rift, within which all the other early hominid discoveries north of the equator had been made. Back in the 1980s, Yves Coppens had made quite a splash with his "East Side Story" that neatly correlated the adoption of bipedality by the earliest hominids with environmental change. The notion was that, in the late Miocene, the doming-up of the African landscape along the north-south axis of the Rift had placed eastern Africa in the rain shadow of the west. As a result, western Africa remained clothed in humid forest that teemed with apes, while the area to the east of the Rift gradually dried out, creating woodlands, bushlands, and grasslands that the emerging hominids negotiated on two feet. The geographical origin of *Sahelanthropus* flatly contradicted this beautiful idea, which had already been battered in 1995 by another hominid discovery in Chad. This was of a 3.6-million-year-old australopith jaw that was given the name of *Australopithecus bahrelghazali*, but that many compared to *A. afarensis*.

Sahelanthropus itself caused a lot of head scratching at first, combining as it did what looked like an "advanced" flat face with a very tiny brain volume of about 350 ml. A computer reassembly of the fractured and displaced bits of the skull later revealed that the face had originally projected rather more than had been evident at first. But it also confirmed the suggestion that the foramen magnum had been positioned underneath the skull, strongly suggesting bipedality. What's more, although the upper canine had been quite pointy, the dentition was not out of line with what might have been expected for an early hominid. None of this, of course, prevented Pickford and Senut from making unflattering comparisons with gorillas, or the Middle Awash group from effectively ignoring *Sahelanthropus* in its discussion of hominid origins.

THE BOURI PENINSULA

But then, the Middle Awash folks had plenty else on their hands. After scouring Aramis, White's team began surveying a bluff known as the Bouri peninsula that lay to its south. Sediments of three distinct ages were exposed here, the oldest of them dating from around 2.5 million years ago. In 1999 the group reported finding a partial skull and some other hominid bones in these lower sediments, and gave the skull the new name of *Australopithecus garhi*. Whether this specimen in fact represents a species distinct from the very similar and slightly earlier fossils from Hadar has been disputed; whatever the case, its most intriguing aspect was archaeological. This was because the ancient sediments that yielded the hominid also produced mammal bones that bore cut marks of the kind left by sharp stone flakes during the dismemberment of carcasses, as well as marks showing pounding as if for marrow extraction. This all added up to clear evidence of butchery, even in the absence of a smoking gun in the form of the stone tools themselves. And the only evident butcher—and presumed meat eater— was that australopith. Moreover, Bouri was more or less contemporaneous with the nearby site of Gona, with its abundant Oldowan tools. Putting everything together, the Berkeley group strongly hinted that nobody could any longer see stone toolmaking as a defining characteristic of our genus *Homo*—even though, largely for philosophical reasons, they favored the idea that their new hominid was ancestral to our genus.

Fossils actually belonging to *Homo* were found higher in the Bouri section, at a million-year-old locality called Daka. Predictably assigned to *Homo erectus* in a 1992 publication, a skullcap from Daka actually appears quite different from anything known from Java; though, since to the best of my knowledge it has been unavailable to anyone but the Berkeley group, it is hard to know exactly what to make of it. Not that this is for want of trying. In 2002 my colleague Jeffrey Schwartz and I visited the National Museum of Ethiopia in Addis Ababa, armed with a letter of permission from the Ministry of Culture to see the Daka specimen and other published Middle Awash hominids. But the technician in charge of the safe containing the fossils refused to open it, and none of his superiors—including the most senior administrators—dared countermand him. Apparently, his refusal had some very influential backing. Still, using just the data published by White's group, the University of Rome's Giorgio Manzi and two Italian colleagues were able to document in 2003 that the Daka skull and other African specimens were quite unlike eastern Asian *H. erectus*. From photos, the Daka cranium also looks significantly different from the penecontemporaneous Buia skull from the Eritrean section of the Afar to the north and east of the Middle Awash; but nobody has yet had a chance to compare the two.

Yet higher in the Bouri sediment pile, near the village of Herto, the Middle Awash team also discovered the second-oldest *Homo sapiens* fossils ever discovered (the oldest being the Omo Kibish specimen discovered by Richard Leakey 30 years earlier, and re-dated in 2005 to around 195,000 years ago). Some 160,000 years old, or only a bit less, the Herto fossils consisted of two adult crania, one fairly complete, and the cranium of an infant. White and his colleagues unsurprisingly found that these were distinct from the European Neanderthals with which they were contemporaneous; and to signify that on a more subtle level they differed more or less equally from all other modern human populations, they awarded them their own subspecies: *H. sapiens idaltu* (from the Afar word for "elder"). This subspecies was the "probable immediate ancestor[s] of anatomically modern humans." The Middle Awash group also emphasized that the

Herto hominids supplied conclusive evidence that humans had evolved in Africa, controverting Milford Wolpoff's Weidenreich-derived notion of "multiregional" evolution—more on that later.

Curiously, the child's skull turned out to bear cut marks—similar to ones White had discovered on the Bodo cranium—which suggested it had been defleshed after death. It also showed an odd polish that might have resulted from its being carried around in a bag, possibly for ritual reasons. Ritual behaviors are characteristic of behaviorally modern humans, but at the same time the archaeological associations of the Herto hominids showed no signs of any behavioral change relative to their predecessors. Indeed, in a nod to the deep past, the Herto artifacts include the latest handaxes known in Africa.

DIKIKA

The Hadar–Middle Awash duopoly in the Ethiopian Afar was broken in 1999 when Zeray Alemseged, now of the California Academy of Sciences, began work at Dikika, an area of badlands across the Awash River from Hadar. The sediments exposed are more or less equivalent to those on the other side of the valley, and sample a similar series of forest-to-woodland-to-bushland environments through which the meandering distributaries of a large river had run into an ancient lake. During the second year of fieldwork, an amazingly well-preserved skeleton of an *Australopithecus afarensis* child was discovered. Nicknamed "Selam" ("peace" in the Afar language), and dated to 3.3 million years ago (120,000 years before Lucy), the tiny girl had died at the age of just over three, as determined using a sophisticated variant of the approach used to age the teeth of the Nariokotome Boy. She had apparently toddled bipedally, but her shoulder blades resembled those of a gorilla much more than those of a human. Young gorillas spend a great deal of time climbing, and careful study showed that Selam had, without doubt, climbed a lot as well. Here was dramatic substantiation of the inference from adult bones that *A. afarensis* had lived a life distributed between the trees and terra firma. Selam also, very unusually, preserved a hyoid bone, the only durable bit of the laryngeal apparatus. This looked very much like its counterpart in an ape, adding to the evidence for apelike vocal abilities in her species.

The archaeology of Dikika also had its surprises. All the sediments exposed there were significantly earlier than the ones at Gona, which had produced the earliest stone tools, some 2.6 million years old. So the Dikika researchers were amazed to find some 3.4-million-year-old mammal bones that were cut-marked, strongly suggesting their possessors had been butchered using sharp stone flakes. The researchers have scoured the exposures looking for stone tools similar to those found at Gona, but as yet have had no luck finding any. One suggestion is that *Australopithecus afarensis* at Dikika may have done its butchery using sharp stone flakes that had been naturally fractured as they were tumbled along riverbeds. Another is that the apparent cut marks were made when sharp-hoofed animals trampled the bones as they lay on the ground, although the Dikika researchers believe they have ruled this possibility out.

The discovery at Dikika naturally spurred the IHO team across the river at Hadar to go back to their mammal bone collections and carefully reexamine them for cut marks. So far they have not found any, but there is nonetheless another line of evidence supporting the possibility that both the Dikika and Hadar australopiths were consuming mammal carcasses well before we have any demonstrable stone tools.

For some time now, paleoanthropologists have been taking advantage of the fact that what you consume leaves a chemical signal in your hard tissues—the teeth and the bones that may, with luck, survive to become fossilized. In the case of the australopiths, the focus has been on the ratios of stable isotopes of carbon that reflect whether you are a browser or frugivore (eating the products of plants, such as most trees, bushes, and shrubs, that use what is known as the C_3 metabolic pathway) or a grazer (eating C_4 plants, such as tropical grasses and some sedges). If you are a hominid and the isotopes indicate, improbably, that you were a grazer, your C_4 signal will most likely have come from grazers you have eaten.

Early studies on South African australopiths yielded a surprisingly high C_4 signal, suggesting that even in the absence of stone tools these hominids were at the very least omnivorous. To vegetable products they would have added such foods as lizards, insects, arthropods, and young antelope. In contrast, apes living in similarly open environments today continue to display a C_3 signature. More recently, studies on eastern African hominid fossils have indicated that a significant dietary shift away from C_3 foods took place after about 4 million years ago. In particular, at both Hadar and Dikika analysis of *Australopithecus afarensis* fossils suggests a diet with a substantial C_4 component. The signal was pretty variable among samples, but on average it didn't change over time, suggesting that A. *afarensis* may have been the hominid that took the plunge away from an ancestral diet of foods typical of forests, and toward resources found in open environments. To what extent this dietary shift reflected active hunting, or the gathering of small prey such as arthropods, or scavenging, or a combination of all of these, is still unclear; but something plainly happened, and it may be what we see reflected in the cut marks on those Dikika bones.

Another independent foray into the Afar was made, beginning in 2004, by Johannes Haile-Selassie, now of the Cleveland Museum of Natural History. In 2005 his team discovered the postcranial skeleton of a large individual, attributed to *Australopithecus afarensis*, in the Woranso-Mille region to the east of Hadar. Dating from about 3.6 million years ago, this presumed male once again drew attention to the large individual size differences present in A. *afarensis*. It also demonstrated a basic consistency in body proportions, with a hind limb that was longer compared to the forelimb than in apes, but shorter than in modern humans. Perhaps predictably in historical context, the Haile-Selassie group downplayed the climbing features present in the upper thorax and forelimb, but although the critical part of the Woranso shoulder blade appears to be broken off, this element seems to confirm the climbing proclivities also suggested by the Dikika group.

THE HOMINID OF THE LAKE AND FLAT-FACED MAN

While all this was going on, important discoveries were also being made in Kenya. Back in the 1960s, the Harvard paleontologist Bryan Patterson had discovered the upper part of a hominid elbow, at a place called Kanapoi, a little to the southwest of Lake Turkana's southern tip. It was thought to be about 4 million years old. For a long time this collecting area lay fallow, until in the mid-1990s Meave Leakey, Richard's wife, turned her attention toward Kanapoi and to deposits of approximately equivalent age at Allia Bay on the eastern shore of Lake Turkana. In 1995 she and her collaborators assigned a handful of hominid specimens collected at both sites to a new species, *Australopithecus anamensis* (anam means "lake" in the Turkana language). Most of the specimens were isolated teeth, but among them were reasonably well-preserved upper and lower jaws and, very importantly, the upper and lower portions of a tibia. These were much more like A. *afarensis* than anything then known from *Ardipithecus*, at that point the ancestral hominid du jour; Leakey and her collaborators suggested that their new species was plausibly ancestral to the former. This suggestion was supported by more precise dating of the *A. anamensis* fossils to between 3.9 and 4.2 million years. Three years later, with more fossils in hand, the Leakey group confirmed their new species was "more primitive" than *A. afarensis,* and in good linear fashion hinted that it might have occupied an intermediate place in the hominid family tree lying between the 4.4-million-year-old *Ardipithecus* and the later *A. afarensis*—a suggestion they later backed away from.

Most interesting about *Australopithecus anamensis,* however, was that it provided the best evidence available for early hominid bipedality. The part of the tibia involved in the ankle joint was particularly instructive, showing as it did that the weight of the body had been transmitted directly down to the foot when the heel struck the ground. The knees had thus passed close together during walking. There also seemed to be evidence that the disparity between male and female canine size was strongly reduced: a hominid feature on which Tim White, in particular, had laid stress. And as if not to be outdone, in 2006 the Middle Awash group duly produced their own fossils of *A. anamensis.* These came in the form of some fragments from the site of Assa Issie, not far from Aramis, which dates to between 4.1 and 4.2 million years ago. The Assa Issie fossils, it was claimed, clinched the position of *A. anamensis* in the middle of a series leading from *Ardipithecus* to *A. afarensis.* At around the same time, the Hadar group chimed in with its own analysis suggesting that a single hominid lineage had changed in a steady, gradual fashion over the 1.2 million years leading from Kanapoi, to Allia Bay, to Laetoli, and finally to Hadar. The Hadar scientists left the position of *Ardipithecus* in abeyance and pointedly ignored the Assa Issie sample, to which they presumably had no access. And in good minimalist fashion they plumped for a "less . . . bushy" early hominid family tree, while admitting that "evidentiary support" for their notion of continuity was "weakest" in the gap between Allia Bay and Laetoli, which is to say, between *A. anamensis* and *A. afarensis.* What's more, there is some evidence for heterogeneity in both the Kanapoi and Allia Bay samples, let alone among the energetically disputed Hadar hominids.

Still, amid all this linear thinking a crack finally began to appear in the monolithic early hominid façade. In 1986 the Russian anthropologist V. P. Alexeev had applied the name *Pithecanthropus rudolfensis* to the famously

flat-faced ER 1470 cranium that, a decade earlier, had started paleoanthropologists thinking seriously about Louis Leakey's species *Homo habilis*.
Whatever it was, though, 1470 was seriously different from anything found
at Olduvai, and in 1992 Bernard Wood, to whom the monographing of the
East Turkana hominids had fallen, finally found himself in a position to
point this out. If he was correct, 1470 needed its own name; and by the rules
of nomenclature, Alexeev's had to take precedence. In this way, ambiguous and unsatisfactory as it might have been, the rather poorly preserved
East Turkana specimen became the standard-bearer of the new kid on the
block, *Homo rudolfensis*. Gradually this idea gained traction, and the new
species gained some additional members. These included a lower jaw found
in Malawi in 1991–92 that dated back almost to 2.5 million years ago, and
a 1.8-million-year-old palate that was discovered in Bed I of Olduvai Gorge
in 1995 (although it was not published until eight years later).

There matters rested until 2001, when Meave Leakey's team (curiously
enough, basically the same group that would contortedly argue six years
later that both *Homo erectus* and *H. habilis* had been present in East Turkana at 1.5 million years ago) bravely broke the paleoanthropological taboo
on creating new hominid genera. During the 1998–99 field season in West
Turkana, new finds indicated to them that something truly unusual had
been present around the lake at around 3.5 million years ago. The principal
specimen concerned, KNM-WT 40000, was a fairly complete and short-
faced hominid cranium that was extensively cracked and infiltrated with
sediment. But it was well enough preserved to show that it was unlike anything else known from the same time frame. So distinctive was it, indeed,
that the researchers felt compelled to assign it to a new genus and species,
Kenyanthropus platyops (flat-faced man of Kenya). Cramming it into an existing genus did not seem to be an option, even for a group of paleoanthropologists playing by the curious rules of paleoanthropology. Further, when
the team compared their new cranium to other hominid fossils, its closest
affinities appeared to lie with the equally enigmatic but very much younger
ER 1470. Accordingly, the scientists suggested that the latter should now be
styled *K. rudolfensis*.

Predictably enough, none of this sat well with Tim White, who responded by invoking Ernst Mayr's complaint that hominid taxonomy was
a "bewildering diversity of names" (which had certainly been the case
when Mayr wrote in 1950, but hardly was now). White even lamented that

hominid fossils had been given such implicitly extraneous names as his own *Ardipithecus ramidus* and *Australopithecus garhi,* and proceeded to excoriate Leakey and her colleagues on the grounds that KNM-WT 40000 was far too distorted by infiltrating sediments to interpret. But in reality, while the cranium shows a little distortion in some areas (and a loss of detail in all), it retains its basic form in most essential respects. Poorly preserved as it is, it presents us with a reality that needs to be confronted in systematic terms, rather than to be hectored away with colorful rhetoric and accusations of "typology" and "populist zeal." After an initial bout of indigestion, most paleoanthropologists seem now to have accepted the Leakey group's conclusions, at least pending better materials.

THE EXODUS

Right through the late 1980s, it appeared that hominids had been confined to Africa until about a million years ago. But then intimations began appearing that the exodus from the natal continent had begun much earlier. Weirdly old dates of 1.6 and even 1.8 million years were obtained for a couple of *Homo erectus* sites in Java, and some flaked cobbles from a site in Pakistan called Riwat seemed to be at least 1.6 million years old. But the clincher that hominids had spread out of Africa at an unexpectedly early date came from somewhere totally unexpected. Nobody would ever have imagined looking for early hominid fossils beneath a ruined town sitting atop a black basalt bluff in the Caucasus, just east of the Black Sea. But, sacked by the Mongols in the fourteenth century and subsequently abandoned, Dmanisi, a former trading center in western Georgia at which ancient Silk Road routes from Iran and Turkey had converged, was a natural magnet for medieval archaeologists. Excavating a storage pit below one of the houses, those archaeologists were amazed to find fossils poking out of its walls: fossils we now know were almost exactly 1.8 million years old, and harbingers of one of the most amazing series of hominid discoveries ever made.

Over the almost quarter century that has elapsed since 1991, five hominid crania and four mandibles have been recovered at Dmanisi, along with parts of three hominid postcranial skeletons, numerous Oldowan-type stone tools, and an extensive fauna of fossil mammals that includes some cut-marked bones. Four of the crania have well-preserved faces, and are matched with four of the mandibles. The first hominid to be discovered at the site was a small mandible with a beautifully preserved set of teeth. In obeisance to tradition and to its presumed origins, this specimen was initially ascribed to "early African *Homo erectus*." In 2000, two more Dmanisi crania, one with most of its face and one without, were described and compared to Kenyan *Homo ergaster* (basically, the ER 3733 and 3883 crania, and the Nariokotome Boy). Just like the Kenyan fossils against which they were evaluated, the two Dmanisi specimens were actually rather dissimilar from each other, as my colleague Jeffrey Schwartz soon pointed out; but because they were from the same site, the Dmanisi researchers deemed them to represent the same species. Notably, at 780 and 650 ml in cranial volume, both of the Dmanisi individuals had smaller brains than any of the Kenyan specimens.

No sooner had the description of the new crania appeared than an amazing new find showed up nearby. I vividly remember seeing it for the first time in Tbilisi, at the home of Leo Gabunia, the gracious old-time paleontologist who led the investigations at Dmanisi, shortly before his death in May 2001. I don't think he could really believe it, and frankly neither could I. The new fossil was a lower jaw that contrasted hugely with the one that had been found a decade earlier. Where the first one was short and delicately built, this one was huge and deep, with prodigiously long tooth rows. Except for its reduced canines, you might not have been excessively astonished to find it in the late Miocene. Accordingly, it was posthumously published by Gabunia, together with French and Georgian colleagues, as the type of a new species, *Homo georgicus*—which also included the Dmanisi specimens that had been described previously. The large mandible was taken to be that of a male; the ones discovered earlier were considered female. Whether or not the extraordinary new fossil—or any of the others, for that matter— really looked like a plausible member of *Homo,* the distinctiveness of the Dmanisi hominids appeared to have been established.

But not for long. Almost immediately, the Dmanisi group announced the discovery of the remarkably well-preserved cranium, with associated jaw, of a young individual. Once more, this very lightly built and quite short-faced individual, with a small braincase of 600 ml, looked rather different from anything that had been seen before. But the researchers decided that somehow all of the Dmanisi hominids—including the big mandible— could be squeezed back into *Homo erectus* (or maybe *H. ergaster*) as a primitive member of the genus *Homo* that was more like *H. habilis* than anything else known. All in all, the new species from Dmanisi "preserves several affinities with *Homo habilis* and *Homo rudolfensis* . . . foretelling the emergence of *Homo ergaster.*" The next find, made in 2002 and 2004, was of the cranium and mandible of an aged individual that was simply described as similar to the skulls already known from Dmanisi. Most remarkable about this particular skull was that all its upper teeth, and all its lower teeth but one, had been lost well before death. Evidently the individual had survived a substantial amount of time—long enough, certainly, for the empty sockets to heal—without benefit of teeth, suggesting that other group members had rallied around in support over a period of what was probably years. The likelihood of this caregiving was increased by the fact that the Dmanisi hominids lived at a time when a cooling trend had converted their environment from a subtropical to a temperate one, making animals the probable major food source.

The Dmanisi team reviewed all their finds to date in 2006, and affirmed that despite their disparities all (except possibly the big jaw, on which they disagreed) had been sampled from a single population that had lived at Dmanisi about 1.8 million years ago. They interpreted the similarities with *Homo habilis* (whatever that might have meant) as primitive retentions, and concluded that other features placed their fossils with *H. erectus,* as represented both in Kenya and in Java. Overall, they opted to recognize the Dmanisi hominids as a new subspecies, *H. erectus georgicus,* which lay close to the origin of *H. erectus.* Time and continuity, it seems, were paramount.

By the time this review was published the real bombshell had actually already hit, in the form of the discovery in 2005 of the cranium of the individual to whom the large jaw had belonged. This beautifully preserved fossil was not described until eight years later, presumably because it was, and remains, hard to know how to interpret it. It is, quite simply, totally unlike any other hominid that has yet been discovered. It has the small braincase (546 ml) and the strongly protruding face of a gracile australopith. But it isn't an australopith, at least in the sense in which the group is understood at African sites. Though I have only seen it briefly, my impression is that this lovely fossil is very different in anatomical detail from any early hominid known from Africa. Nonetheless, if it is not an australopith, it is yet more unlike any well-established member of *Homo*—even though, for unclear reasons, the Dmanisi team claimed "close morphological affinities with the earliest known *Homo* fossils from Africa."

This brings us directly to the issue of what should be accepted as a member of *Homo,* a question that is less straightforward than you might think. Basically, our genus as commonly recognized started off with one single species—our own—and then "just growed" over the years as more fossils were accreted into it, each new entrant stretching its bounds a little further. As a result, most current notions of what *Homo* contains—and, by implication, of what *Homo* is—have not resulted from any coherent attempt to make sense of the morphological variety seen in the hominid fossil record. The closest thing to such a review was made by Bernard Wood and his former student Mark Collard in 1999, as a follow-up to work originally begun by Wood in 1992. Wood and Collard began with an explicit rejection of the "Man the Toolmaker" criterion for membership in the genus *Homo,* and followed this by correctly pointing out that in principle all genera should be monophyletic. What this means is that the grouping should consist of all the species descended from a particular ancestor. After all, groups defined by descent are the only "real" groups that occur in Nature: anything else is entirely arbitrary. For most systematists, then, descent is the essential criterion for genus membership, and the matter rests there.

But as paleoanthropologists, Wood and Collard were unhappy about adopting the criteria routinely employed for nonhuman taxa. Their main complaint was that, at least in theory, a strictly monophyletic genus might unite species that are differently adapted; and in paleoanthropology, adaptation has traditionally taken massive precedence over descent. The two researchers thus turned away from ancestry and proposed to define the genus as a group "whose members occupy a single adaptive zone." Whatever that might mean: birds, bats, and insects all fly, though this is hardly a useful criterion by which to group them. Still, playing around with a variety of alternative trees for the hominids, Wood and Collard rapidly realized that, whatever else the various fossils often attributed to *Homo habilis* and *H. rudolfensis* might or might not have had in common, they did not group comfortably into an adaptive unit that, in addition to the type species *H. sapiens,* would also include *H. ergaster* and later hominids attributed to the genus *Homo.* They therefore proposed excluding these and other "early *Homo*" fossils from membership in our genus. Instead, *Homo* would basically just include those hominids with bodies and jaws of essentially modern proportions. The morphological variety left within the genus *Homo* under this formula was still very broad, because despite relative postcranial uniformity there is a lot of difference in skull structure between *H. ergaster* and *H. sapiens;* within mammalian genera, skull form typically doesn't vary as much as this. But the reduced grouping favored by Wood and Collard was nonetheless reasonably compact, and since—as far as we know—modern body proportions (possibly unlike bipedality itself) only evolved once, this more limited concept of *Homo* is also conveniently monophyletic.

Wood and Collard's solution left *Homo rudolfensis, H. habilis,* and the various other early *Homo* specimens in a sort of taxonomic limbo—although, with typical paleoanthropological insouciance, their preferred solution to this problem was to make the genus *Australopithecus* even untidier by brushing these fossils into it. But the countervailing advantage was to make the genus *Homo* a much more readily definable unit, thereby establishing a fairly practical yardstick by which to evaluate the putative membership in *Homo* of such new fossils as those from Dmanisi. And the Georgian fossils clearly fail this more stringent test, as Wood and Collard themselves have recently pointed out. These hominids had possessed tiny brains, and whether or not they actually belonged to a single species, they had the overall skull proportions of australopiths. What's more, a partial skeleton (which might belong with the new skull) and the postcranial bones of three other individuals suggest that, unlike the Nariokotome Boy (and other individuals from Turkana represented by individual bones), the Dmanisi hominids had been very short in stature. Yet they are also said to have had basically modern proportions, while at the same time showing "primitive features." The last word has evidently to be written on what exactly the Dmanisi postcrania tell us about the emergence of the modern body.

The last word has also to be written on whether all the Dmanisi fossils should be attributed to the same species. Even before the fifth skull was discovered, some researchers had suspected heterogeneity in the assemblage; with the new cranium it became a no-brainer. The principal—indeed the only—reason for lumping the Dmanisi hominids together is that they all come from the same site. If they had been found at sites miles apart, or at different stratigraphic levels, nobody would have had any trouble recognizing that they were diverse. Yet the geologists have concluded that the Dmanisi deposits may have accumulated over several hundred years. This lapse would have allowed plenty of time for mobile hominids of more than one kind to visit this evidently very favored place, possibly a lakeside around which plants and animals abounded, including the predators that may have contributed to the hominid bone accumulation. That multiple hominids may have visited such a place is hardly surprising: after all, there is evidence for at least four different kinds of hominid sharing the landscape around Lake Turkana at around 2 million years ago.

As each subsequent specimen came to light, the Dmanisi researchers became more insistent on abandoning the earlier notion of *Homo georgicus* and allocating their very heterogeneous assortment of hominid fossils to the single species *H. erectus*—a species based, as you'll recall, on fossils from Java with which none of them shared any significant derived similarities. When describing the amazing fifth skull in 2013, the team justified this allocation by observing that "variation in Plio-Pleistocene fossil hominids tends to be misinterpreted as species diversity." And they accordingly proceeded to attribute the entire Dmanisi assemblage to the subspecies *H. erectus ergaster*—which of course also included all of the Kenyan "early African *erectus*" fossils as well. This in itself was a stretch, not only because a huge morphological variety was involved, but also for the opposite reason: among living primates it is often hard, if not impossible, to tell members of different subspecies of the same species apart purely on skull anatomy. Still, even this was not the limit. The Dmanisi group went further yet, creating a previously unheard-of taxon, the sub-subspecies—*H. erectus ergaster georgicus*—to contain the whole hominid assemblage from their site.

Faced with an array of morphologies like that found at Dmanisi, only paleoanthropologists—gloriously isolated from mainstream systematics, careless of the established rules of nomenclature, and still in thrall to Mayr's midcentury fundamentalism—could possibly ever have made a suggestion like this. If the variety of morphologies seen among the Dmanisi hominids represents no more than epiphenomenal variation within a single subspecies, then we might as well abandon morphology as a systematic criterion. And since we have already had to acknowledge that neither a fossil's locality nor its age has any necessary bearing on what it is, what are we left with? Literally, we would have nowhere to go in trying to sort out the rich diversity that exists at lower ends of the taxonomic hierarchy as well as among larger taxa. What a nihilistic prospect! We must wait to see how the paleoanthropological community responds to the Dmanisi team's suggestions. Meanwhile, what better evidence could one seek that, in some quarters at least, the paleoanthropological mind-set continues to be dominated by the notion that human evolution was no more than a simple linear struggle from primitiveness to perfection, by a hugely variable and constantly changing species? Or of the associated smug feeling, also inherited directly from Mayr, that it is naïve, unsophisticated, and typological to imagine that manifest anatomical diversity among fossils might actually encode a systematic signal.

CHAPTER 9

MOLECULES AND MIDGETS

BY THE TIME THE 1980s CAME ALONG IT WAS BECOMING PRETTY
evident that, without access to Madagascar, my future as a lemurologist was
going to be pretty limited. So once I'd summarized my thoughts about lemurs in a book that came out in 1982, paleoanthropology beckoned me
back. One obvious thing to do was seek out a potential field site, though
obvious places to look for human fossils were few and far between. Accordingly, while planning my return to New York from studying monkeys on
the Indian Ocean island of Mauritius, my eye alighted on Djibouti. This
tiny former French territory poked into the Ethiopian part of the Afar Triangle, well under a hundred miles to the east of Hadar. And as far as I could
tell from the sketchy geological maps available, the same general series of
sedimentary rocks that had yielded the amazing Afar fossils were exposed
right across into the brand-new Republic of Djibouti, making the place at
least worth a look. Additionally, Djibouti was a refueling stop on the flight
from Mauritius back to Paris, so the decision to go there more or less made
itself. But on the very night I arrived in Djibouti's steamy and decaying port
city, a drunken gang of Yugoslav sailors got out of control in the cheap bar/
hotel/bordello that was all my budget permitted. They rampaged through
the premises in the wee hours, breaking into rooms and robbing and knifing
their occupants. I was one of the very few clients of the hotel who escaped
unhospitalized, following an agonizing eternity spent leaning hard against
the flimsy and violently shaking door of my room, as someone pounded

furiously on it from the outside while screams, thuds, and the sounds of breaking glass reverberated through it.

Call me unmotivated, but this experience dimmed my enthusiasm for fieldwork in Djibouti. What's more, the place eventually turned out to be a little disappointing. Pleistocene fossil faunas were reported a few years later by French researchers, but only one hominid ever turned up, a maxilla fragment that was said to be "more like *Homo sapiens* than *Homo erectus*." My wanderlust later returned, of course, taking me to such places as Yemen (a wild and austerely beautiful place where, at different times, I feared I was being kidnapped by rebel tribesmen, or about to be shot by the army) and Vietnam (where, although the landscapes were ethereal, the fossils were elusive). But the immediate effect of my Djibouti experience was to make me realize that every bit as important as finding new fossils was ensuring that fossils already on museum shelves had been properly interpreted. Since it was already very clear to me that the minimalist approach inherited from Ernst Mayr had resulted in a severe underestimation of the variety present in the human fossil record, and consequently in a misleading perspective on how we actually became human, I started thinking again about what fossil species were, and about how paleontologists should go about recognizing them.

This was not an issue that was much on anyone's mind at the time. After all, almost everybody thought they knew that species were merely arbitrarily defined segments of evolving lineages. Still, this did not mean that everyone agreed on exactly what had happened, and opinion was beginning to divide into two camps. One was occupied by Milford Wolpoff and his colleagues and students, who had cleverly bounced back from their disappointment at East Turkana by formulating a revised version of the Single-Species Hypothesis that rapidly became known as the Multiregional model. Harking back to Franz Weidenreich's prewar ideas, this explained the geographical variety of *Homo sapiens* today by proposing that (1) the major geographical groups of humankind had deep roots in time going back to the days of Java Man and Peking Man, the differences among them having evolved in a state of quasi-isolation; but also that (2) the unity of the overall species had nonetheless been maintained throughout by gene flow between the geographical varieties at the edges of their distributions. Once its proponents realized that, as initially articulated, this sequence of events involved several lineages independently, and impossibly, crossing the species threshold to *H. sapiens,* the problem was neatly solved by extending the concept of *H. sapiens* to embrace *H. erectus* as well. The sclerotic reductionisms of the New Evolutionary Synthesis remained alive and flourishing.

The main competition to multiregionalism, at least as it related to the origin of *Homo sapiens,* came from scientists in England and Germany who were beginning to articulate what has come to be called the Single African Origin model. These researchers saw that, in an admittedly sparse fossil record, the earliest intimations of modern human anatomy were found in Africa. And they soon found themselves receiving support from an unexpected quarter. The techniques of molecular systematics were rapidly being refined, and nowhere more quickly and imaginatively than in the Berkeley lab of Allan Wilson, where Vincent Sarich had worked. The genes, it had long ago transpired, consist physically of sequences of "bases" along the long DNA strand. These bases are of four kinds: A, T, C, and G; strings of these "letters" spell out the inherited genetic instructions on which the building of each new individual is based. In the 1970s, new techniques were coming on line that allowed the direct reading of these instructions, so that attention moved away from the proteins—such as those albumens Sarich and Wilson had studied—that the genes coded for, and toward the structure of the DNA itself.

By comparing the DNA of humans and chimpanzees, Wilson and his student Mary-Claire King were able to show in 1975 that purely structural differences in their DNA—in their protein-coding genes—were too small to explain their large anatomical differences. Those anatomical dissimilarities must thus have been due to differences in how active the genes were during the developmental process. This was a highly prescient and hugely important conclusion that more recent research has entirely borne out: changes in the regulation of gene activity and expression have clearly been key factors in evolution. Where such changes extensively affect developmental processes, the activity of the more abundant "transcription factors" that govern the activity of protein-coding genes almost certainly goes a long way toward explaining many of the anatomical gaps we see in the fossil record. Such gaps probably include the apparently rather abrupt appearance of the basically modern body anatomy evident in the Nariokotome Boy, and the emergence of the anatomically very distinctive *Homo sapiens.*

Soon Wilson and his collaborators were moving forward by focusing on DNA of a particular type. Most of our DNA is contained in the nuclei of our cells, but a small quantity is found in the mitochondrion, one of the numerous kinds of hard-working organelles that lie in the cytoplasm between the nucleus and the enclosing cell membrane, and are responsible for cellular function. There is much less DNA in the mitochondrion than there is in the nucleus (though because there are lots of mitochondria, there are many more copies of it). And for the purposes of molecular systematics this "mtDNA" has another huge advantage. Though most nuclear DNA becomes inconveniently jumbled up between generations when the egg and the sperm combine, mtDNA provides a clear record of heredity because it is passed along uniquely through the maternal line (the father's sperm is, in essence, just a cell nucleus). So any differences in the mtDNA of any pair of individuals or species must be due purely to mutations (which had structurally turned out to be simply spontaneous changes that constantly occurred in the four possible kinds of nucleotides) that had accumulated since they shared a common ancestor.

In 1987 Wilson and his students Rebecca Cann and Mark Stoneking looked at the mtDNA of a sample of 147 people from around the world. Assuming a constant rate of change, they calculated from the differences observed in this sample that our species had originated around 200,000 years ago. What's more, the mtDNA tree they derived indicated an origin of *Homo sapiens* in the continent of Africa—as did the fact that mtDNA diversity was greatest among people of African descent, implying that people had been evolving in Africa longer than elsewhere. The exact date at which *H. sapiens* differentiated has been debated extensively by molecular systematists, but there is no argument among them about its relative recency, or about their later discovery that, at some time subsequent to its origin, the *H. sapiens* population passed through a bottleneck—a substantial population contraction—before definitively leaving the continent of Africa.

I naturally enough liked these findings a lot, not only because they confirmed what I already saw in the fossil record, but also because Wilson and colleagues' molecular approach incorporated the assumption that, even if our species hadn't yet reached its end, it had indeed enjoyed a definable beginning. You couldn't do molecular systematics like this unless you admitted that species somehow had real existences; and this acknowledgment was something that opened a side door, as it were, to a more nuanced appreciation of the human fossil record. But its implications have taken a while to sink in, and I must admit that, at the time, my own major preoccupation was with the general dismissal of the truly distinctive Neanderthals as a mere extinct variety of *Homo sapiens.*

This was because it seemed to me that not only anatomically, but also behaviorally—as reflected in their archaeological record—the Neanderthals were hugely unlike us. I felt that by denying these extinct relatives a discrete identity, paleoanthropologists were passing up the opportunity to understand them, in their own individual terms, as the distinctive beings they most certainly were. Having grown weary of waiting for the Neanderthal experts to come to terms with this, in 1986 I wrote a paper in which I broached the problems attendant on recognizing fossil species via their morphologies. I emphasized that speciation, the process whereby new reproductive entities come about, is quite evidently not simply a passive consequence of morphological change. Some species accumulate a large amount of morphological variety without dividing into daughter species, whereas some populations that can barely (if at all) be told apart by eye totally ignore each other for reproductive purposes. This evident disconnect between morphological distance and species status may not matter much if you are able to observe the living creatures interacting (or not) in the same place; but it is a hugely complicating factor if they live far apart, or if you only have fossil bones to go on.

Nonetheless, practical experience—not least with the lemurs—suggested that as a rule very closely related living primate species do not typically differ hugely in the morphology of their bones and teeth. So, if there are two distinctly different and consistent hard-tissue "morphs" present in an assemblage of fossils that you are trying to sort out, chances are that more than one species is present—as long as you can eliminate sex differences as a potential explanation. What's more, since bony differences among closely related species tend to be subtle, chances are that paleontologists will *under*estimate the number of species represented in a fossil sample. This undoubtedly introduces a significant element of uncertainty, but underestimation is actually preferable to the opposite bias, since it will just oversimplify the evolutionary patterns you infer, rather than distort them. At the same time, this bias also means that it is probably futile as well as unhelpful to try to recognize fossil subspecies—distinctive regional variants that readily arise within species, but that can equally easily disappear through intermixing. Subspecies are the engines of evolutionary innovation in the sense that they provide the platforms for future species;

but until speciation confers historical individuality on them, they remain evolutionary ephemera.

The practical difficulties involved in recognizing closely related species in the fossil record have hardly disappeared since I wrote that 1986 paper. Despite my best efforts I, for one, have certainly never found the elusive silver bullet for spotting them. However, one gratifying side effect of the search for fossil species has been a diminution in recent decades of paleoanthropologists' usage of the term "grade." The grade is a peculiarly paleoanthropological concept that I once felt impelled to describe as one of the most destructive canards the field had ever seen fit to inflict on itself. This is because these undefined groupings of fossils that look vaguely similar (have around the same brain size, for instance) are the products of human minds, and not of Nature. They have no independent existence; they are merely a heuristic device that is nonetheless often mistaken for something real. By itself this might not be so bad, but thinking in terms of grades also allows you to ignore the nested sets that Nature actually produces. Saying that hominids passed through a *Homo erectus* grade" in their evolution, for example, comes close to implying that there is no pattern of ancestry and descent in human evolution. Instead, inexorable natural selection has done all the heavy lifting, and we have no reason to inquire any further. But if we were accordingly to stop there, we would miss the important fact that, over the last couple of million years, brain enlargement with time in the genus *Homo* has taken place *independently* within at least three lineages: *Homo erectus* (as defined in eastern Asia), and those leading in Europe to Neanderthals and in Africa to *H. sapiens*. This really is significant, because if this vaunted proclivity is thus a property of the diverse genus *Homo* as a whole, and not specifically of our own lineage, then we have to seek its roots in a very different place.

Still, back in 1986 my immediate concern was to point out that modern *Homo sapiens* is in many ways extremely distinctive, and that only a very few fossils (all of them relatively recent) fit comfortably into the morphological envelope described by living populations of our species. There isn't much in the Pleistocene fossil record that shows the balloon-like braincase, with a very small face retracted right beneath it, that all modern *H. sapiens* exhibit. Equally distinctive was *H. neanderthalensis*, with a long list of derived features that then appeared to be shared with little else (its close relative from the Sima de los Huesos had yet to be discovered). There was no doubt in my

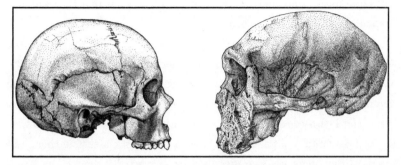

Contrast in cranial construction between a modern Homo sapiens (from El Hesa, Egypt, left) and a Homo neanderthalensis (from the Grotta Guattari, Monte Circeo, Italy). To scale. Drawn by Don McGranaghan (left) and Diana Salles (right).

mind that *H. sapiens* and *H. neanderthalensis* were two individuated entities, each of which was well launched on its own independent evolutionary trajectory by the time the two came into contact. And it seemed to me that we were denying that individuality by bundling the Neanderthals into *H. sapiens,* as the bizarre—and by implication inferior—subspecies *H. sapiens neanderthalensis.* Yet some paleoanthropologists were—and still are—hell-bent on doing this: there is a lively school of thought that sees Neanderthal-modern hybrids everywhere in the late Pleistocene. Indeed, it displays a quasireligious fervor. I distinctly remember some time ago feeling like a heretic pursued by the Inquisition when I was pilloried by a colleague all over the Internet for penning what I thought was an extremely diplomatically worded commentary urging caution over interpreting a newly found, 24,500-year-old juvenile skeleton as the product of Neanderthal-modern hybridization. Yet the child had lived at least 200 generations after the last anatomically recognizable Neanderthal had gone extinct!

As distressing as the denial of a distinctive Neanderthal identity was the fact that cramming them into *Homo sapiens* made a mockery of our own anatomical uniqueness, a distinctiveness that was just as effectively disguised by the large majority of Anglophone paleoanthropologists who classified such crania as those from Kabwe, Petralona, Arago, and Bodo as "archaic *Homo sapiens.*" In implicit acknowledgment of how different from us such fossils actually were, non-English speakers at the time mainly preferred to allocate these fossils to *H. erectus.* But nobody saw fit to argue this point much, since almost everyone thought anyway that *H. erectus* had gradually given rise to *H. sapiens.*

My strong alternative preference was to place the Kabwe-Petralona-Arago-Bodo group in *Homo heidelbergensis,* the species defined by Otto Schoetensack's Mauer jaw (which, it later transpired, had mandibular and dental features that were neatly matched in jaws from Arago). Even then, some later Pleistocene skulls such as the ones from Laetoli and Ndutu were left in taxonomic limbo, although the difficulty of classifying them emphasized that there was significant taxonomic variety among the hominids known from this time. However you sliced it, the basic fossil signal was clearly one of diversity. Which, as I pointed out, was precisely what one would expect given the unsettled environmental conditions that had reigned over the Pleistocene. Frequent climatic changes would have regularly fragmented hominid populations, providing optimal conditions in which evolutionary novelties could arise and become fixed in the resulting small isolates. Reunification once the climate ameliorated would have allowed those isolates to expand again, resulting in competition and triage.

BLAME THE ENVIRONMENT

Accurately documenting the environmental ups and downs that were almost certainly responsible for rapid hominid evolution over the Pleistocene has been made possible by means of measuring and calibrating them that were developed over the second half of the twentieth century. Most importantly, these involve measuring the ratios of heavier to lighter oxygen isotopes in cores drilled in the muds of the ocean floor and in the Antarctic and Greenland ice caps. Both kinds of core contain continuous and datable records of this type, and are good proxies for polar ice cap volume and thus for global temperatures. This information allows for an infinitely more precise appreciation of past climates than traditional glacial geology permitted.

The study of these cores reveals that, while world climates had been irregularly deteriorating for a long time, a significant episode of global cooling began a little before 2.5 million years ago. This is now officially taken as the starting point of the Pleistocene epoch (although many are nostalgic for the 1.8-million-year mark at which it used to be defined). After this point, average global temperatures shuttled from warmer to colder and back on a pretty regular 41,000-year cycle related to cyclical changes in the tilt of

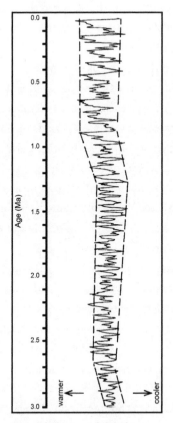

*Simplified curve of global
temperature changes over the
past three million years, as
inferred from deep-sea core
oxygen isotope ratios. See text.
Drawn by Jennifer Steffey.*

the Earth's axis. However, a little before a million years ago, the ongoing
cooling of the oceans under the continuing influence of the vast Antarctic
ice cap began to nudge world climates toward the familiar "Ice Ages" cycle,
in which the northern polar ice cap advanced and retreated over northern
Eurasia and North America about every 100,000 years. In Africa, the northern glacial advances were reflected both in lowered average rainfall and in
greater climatic variability. But most significantly for the ancient hominids,
those grand climatic curves were not smooth. Each major excursion, from
maximum glacial cold to greatest interglacial warmth, was interrupted by

innumerable smaller oscillations that ensured the later evolution of the genus *Homo* would take place against a backdrop of supremely unsettled climatic and environmental conditions.

The great lesson of the cores—as of numerous studies on land—is that short-term fluctuations in the external environment have been very much the norm in hominid experience, even as technological traditions seem typically to have lingered for extended periods of time. In weighing their potential effects—which of course exerted themselves with a fine disregard for hominid convenience and physical adaptation—it is important to remember that climatic excursions had geographic as well as environmental consequences. Not only did temperature and humidity changes cause vegetation zones to move south and north, uphill and down, but coastlines changed with falling and rising sea levels as evaporating seawater, falling as rain or snow, was locked up in the polar ice caps during cold times and released during warmer ones. What was an island at one moment might have become a peninsula a mere few thousand years later.

All in all, the Pleistocene world was one that might have been expressly designed to promote change in the biota. In Africa, climates shuttled between wetter and drier conditions. Most notably, incremental decreases in humidity near 2.8, 1.7, and 1.0 million years ago all translated into the spread of more open habitats and significant changes in the fauna, although once faunas had been "winterized" by such events the overall emergence of new species and the extinction of old ones generally subsided to a background level, at least until the end of the Pleistocene. Those dates also broadly coincide with important inflection points for hominids: at 2.5 million years ago, the first stone tools marked a significant cognitive leap, while at around 1.8 million years ago the modern body form appeared and brain sizes began their remarkable increase.

Subsequently, though, the hominid dynamic diverged from the norm. In the period following about 2 million years ago the human lineage began to exhibit a substantially greater rate of evolutionary innovation than any other mammal group, and the result is that human beings today are vastly more unlike their predecessors of 2 million years ago than any other organism you might care to name. So why this accelerated modification, over a period when other mammals were basically going about business as usual? Clearly, the unsettled environmental conditions that constantly buffeted hominid populations would have provided the essential background; but there was equally clearly some additional ingredient that impacted only hominids. That extra element was most likely material culture. For the ability to modify the circumstances of life that culture conferred was exclusive to hominids, and might have affected their rates of innovation in one of two ways.

First, as time passed and projectile weapons became more sophisticated, hominids became their own most dangerous predators. The possession of remotely killing weaponry would have placed hominid groups in an unprecedented kind of competition with each other, perhaps building on behavioral proclivities that resembled the observed propensity of environmentally stressed wild chimpanzees to indulge in organized conflict with neighboring groups. This kind of antagonistic interaction might have produced strong—and ongoing—intergroup selection for wilier behavioral strategies, as well as for any cognitive or physical attributes that might have been associated with them. But it is hard to imagine the archaeological record improving any time soon to a point at which this possibility might be adequately tested.

Second, it is already a pretty good bet that the possession of material culture—fire, perhaps, and certainly the ability to fabricate clothing and shelters—would have allowed hominid populations to expand farther in good times than they could otherwise have done. The downside of such overexpansion was that these populations would have become more vulnerable to geographical and demographic contraction when environmental conditions deteriorated, as they inevitably did. This would have multiplied the isolating effects of fragmentation, and in this way material culture could well have leveraged the potential of climatic instability to spur population replacement and evolutionary change among the hominids.

BACK TO THE BEGINNING

Whatever the underlying cause, it is almost certainly significant that hominid brains apparently began to expand only well after our fossil relatives finally committed themselves to life in the open. Before this point, the pattern of hominid evolution seems to have been pretty typical of that of any successful mammal group. It was only much later, once basically modern body form was on the scene, that the brain-enlargement trend picked up. So what about those early bipeds that eventually invented stone cutting tools? As we saw, Raymond Dart and his followers correlated the very first adoption of bipedality with a predatory—nay, bloodthirsty—lifeway that both anticipated and drove everything that was to follow in human evolution. More sober assessments have subsequently suggested something different. As noted, those cave accumulations of broken hominid and other bones at the early South African sites seem to have resulted largely from the activities not of mighty hunters but of humble mammalian scavengers—something that fits well with the hazardous lives the australopiths must have led in their new surroundings. Though there is precious little to indicate that these human precursors were the vicious hunters envisaged by Dart, the expanding African woodlands and grasslands in which they lived teemed with fearsome carnivores that would have taken a serious interest in slow-moving bipedal primates.

So to what should we compare those early precursors? The great apes to which the australopiths have most often been likened are basically forest creatures, and with today's depleted fauna they have few serious predators anywhere (except for *Homo sapiens*). Also they tend not to modify their habits much even when, as some chimpanzees do, they occupy more open environments. But predation would have been a huge factor in the ancient hominids' newly adopted habitats, and the experimental bent they showed once they were out there contrasts strikingly with the dietary conservatism of chimpanzees. Accordingly, my colleagues Donna Hart and Bob Sussman have suggested that a better comparison for the early hominids might be with dedicated open-country primate relatives such as baboons—which were, as you'll recall, the basis for the original primate model developed by Clifford Jolly in the 1960s.

Far from living in the smaller groups that are typical of apes (which are more comparable to human extended families in this respect), savanna baboons and the ecologically equivalent macaques of Asia live in very large groups of a hundred individuals or more. These hordes move across the landscape in spatial arrangements that protect the most reproductively valuable individuals at their core; and within each highly structured group there is constant competition for status and, among males, for access to reproductively active females. Baboons and macaques do not show the conspicuous cooperativeness ("prosociality") of modern humans: something that is, moreover, only mildly foreshadowed in apes. And these open-country monkeys suffer very high rates of mortality from predation, despite refuging at night in trees and on inaccessible rock faces. In short, baboons and macaques exhibit the social characteristics not of predators, but of prey animals. Hart and Sussman's suggestion is that this would also have been true of the small-bodied, vulnerable, and significantly slower australopiths. From their fossil associations, we can be pretty certain that the early hominids hewed to the baboon principle of keeping their options open, dividing their time between more dangerous open country and safer forests. And it seems at least a good bet that they correspondingly lived in large groups that were structured to protect core individuals at the expense of the more dispensable. If this were indeed the case, they were very unlike their remote modern descendants in their social organization and proclivities.

Still, the fact that stone tool manufacture and butchery were invented by "bipedal apes" suggests that significant cognitive change had begun before *Homo* arrived on the scene, and it's unfortunate that we currently have no way to know just how the amazing transition among hominids from prey to predator was accomplished, and how long it took. The fossil record doesn't yield much that is of direct relevance, and in any event there can be little doubt that much of this mind-boggling shift was achieved through behavioral rather than anatomical adjustment. But here we are frankly guessing, and the guesswork is not easy. For example, it has been proposed that the shift toward dietary dependence on animal proteins was accomplished via "power-scavenging," in which individuals cooperated to drive predators off carcasses by hurling stones at them. If so, the early hominids must have been significantly different from chimpanzees, which, strong as they are, cannot throw objects very far. The fact that some early australopiths, at least, retained very apelike shoulder anatomies might suggest that they were not very good throwers, either—in which case, we would probably have to conclude that power-scavenging was adopted at a later stage. But since we entirely lack any living model to consult, we are entirely in the realm of conjecture here.

Our knowledge of the australopiths was considerably expanded in 2010 with the description of a new gracile species, *Australopithecus sediba*, from the collapsed cave site of Malapa, not far from Johannesburg. Almost exactly 2 million years old, *A. sediba* has been described as intermediate between other gracile australopiths and *Homo*, though some would prefer to say that it shows how *Homo*-like an australopith can be—mainly in its pelvis, which nods toward later hominids in being a bit narrower and less aggressively flaring at its top than its counterparts from Hadar and Sterkfontein.

To complicate matters, my colleague Yoel Rak has even argued that there are significant distinctions between the two partial skeletons from which *A. sediba* is mainly known: differences that suggest more than one taxon might be present. Still, the one reasonably complete cranium known has an apelike volume of around 420 ml, and though the teeth are not by any means massive, there is little to suggest that the skull of *A. sediba* was substantially "advanced" relative to other gracile australopiths.

Until paleoanthropologists have fully digested the wealth of fossils from Malapa—which, in a hugely gratifying break from mercantilist paleoanthropological tradition, are already freely available to researchers—the main message of *Australopithecus sediba* is to emphasize just how diverse hominids were during the late Pliocene and early Pleistocene: at least half a dozen different species, in three distinct lineages, are known to have populated the Old World in the same general time bracket as *A. sediba*. This diversity was not new, of course: it reflected a pattern that had apparently existed from the family's earliest days, and would continue to characterize it during the tenure of the genus *Homo*. But the recognition of the new species did represent a gradually evolving standard among paleoanthropologists. It remained impermissible to name new taxa on the basis of fossils already known; but this could now be contemplated for newly discovered fossils.

THE ODD LITTLE GUY FROM FLORES

No mention of diversity among hominids would ever be complete without some mention of the strange fossils found in 2003 and 2004 during excavations at the Liang Bua Cave on the Indonesian island of Flores. For no hominid fossil find—not even the fraudulent Piltdown—had ever come from farther out in left field than this strange phenomenon, and I remember feeling totally bowled over by my first sight of casts of it.

A pile of deposits over 30 feet deep in the floor of Liang Bua yielded a series of hominid remains that were scattered among levels between about 12,000 and 80,000 years old. The most striking find was of the partial skeleton of an individual (dubbed LB1) that had lived about 18,000 years ago. This was pretty recent, well within the period during which modern human beings had been around in the Southeast Asian islands. But the skeleton itself was like nothing anyone had ever seen before. Although adult, LB1 had stood no more than about three and a half feet tall, and he (or maybe she) turned out to have been very oddly proportioned, with, for example, a poorly arched foot that was extraordinarily long in proportion to the lower leg; a pelvis with a sideways flare many compared to australopiths; and wrists and shoulders that were either somewhat apelike, or as one might have expected to find in a much more ancient hominid. The skull was equally curious. To go along with its diminutive body, LB1 had possessed a tiny brain of no more than about 380 ml, smaller than those of almost all australopiths known. But its face was much tinier, more lightly built, and less protruding than that of any australopith, and its molar teeth were very small.

What to make of this strange and oddly recent phenomenon? Liang Bua had yielded something that, except in stature, was hugely unlike any australopith known. Yet except in its very small (and badly worn) molar teeth, it was also drastically dissimilar to *Homo sapiens*—or, indeed, to any other species plausibly placed in *Homo*. The Australian and Indonesian scientists who described LB1 considered various explanations for the individual's singular morphology, and arrived at the conclusion that it was a "dwarfed" descendant of *H. erectus*, long known from the island of Java to the west. In principle, this might not have been implausible; species isolated on islands often become reduced in body size, and in the same deposits that yielded the hominids the researchers also found the remains of diminutive elephant relatives. But there really is nothing in the anatomy of LB1 to suggest any particular affinity with *H. erectus*, and it has also been argued that any known dwarfing process would have left the Flores form with a much larger brain than it actually has. Moreover, it is hard to see how reducing your body size could have been at all advantageous in an island that in Pleistocene times was crawling with Komodo dragons, fearsome lizards that grow up to ten feet long.

One alternative to dwarfing is pathology. In the spirit of the Rickety Cossack, numerous authors have suggested that the LB1 skeleton is that of a *Homo sapiens* (or conceivably a *H. erectus*) that had suffered from a disease or a developmental anomaly that led to its curiously stunted anatomy. But none of the medically documented modern human syndromes suggested quite fits the bill. The fossil bones themselves do not show any anatomical or structural indications of pathological development or degeneration, and a very similar mandible of a second individual, together with various other assorted bones, suggests that LB1 was a representative member of a hominid population that occupied Liang Bua over many tens of thousands of years.

These various considerations leave us with only one other obvious possibility, namely that the Liang Bua population was descended from a hominid lineage that had never been tall and had retained the anatomical heritage of an extremely archaic ancestor. Indeed, a sufficiently archaic ancestor might well have had to disperse out of Africa before even the precursors of the Dmanisi hominids. How and when such an early wanderer ever contrived to reach the island of Flores, permanently isolated from any nearby land mass even at low sea levels, remains a mystery. But then again, so does the presence on the island of simple stone tools that may be up to a million years old. These are certainly indications of early hominid presence in Flores, though what kind of human made them is anybody's guess.

Perhaps the greatest mystery of all is why the scientists who described LB1 elected to classify it as *Homo floresiensis*. If it wasn't pathological, then it obviously belonged to a previously unknown species; but there is little about the skeleton to support its assignment to the genus *Homo*. Once again, it seems, the idiosyncratic paleoanthropological mind-set was at work: if it was a hominid (no argument there), and it was not an *Australopithecus* (which it clearly wasn't), then the only remaining option, however much it strained credulity, was to call it *Homo*.

CHAPTER 10

NEANDERTHALS, DNA, AND CREATIVITY

ONE OF THE TALENTED MOLECULAR GENETICISTS WHO HAD been exposed to the creative genomic work that was the hallmark of Allan Wilson's Berkeley laboratory was a young Swede called Svante Pääbo. Inspired by his experience there, Pääbo followed up in 1997 with the first extraction of DNA from a fossil hominid: a small fragment of the mitochondrial genome, obtained from an arm bone of the original Neanderthal from the Feldhofer cave. This was an amazing technical accomplishment, because DNA is a long and fragile molecule that needs constant repair in life and rapidly degrades after death into shorter and shorter fragments. The inexorable process of degradation is retarded by cool, dry conditions that, along with the fact that mtDNA comes in numerous copies in each cell, explain why there was still some DNA left in the Neanderthal to extract. The long and arduous struggle with technical issues that made this feat possible is lucidly explained by Pääbo in his recent book *Neanderthal Man*.

His success meant that, at last, direct genomic comparison had become possible between *Homo sapiens* and an extinct relative. And the fraction of the circular mtDNA molecule that Pääbo and his group contrived to sequence turned out to fall entirely beyond the range of variation observed among all modern humans. In the part of the mitochondrial genome

involved, individual modern humans on average show eight nucleotide differences, compared to about 55 between humans and chimps. The Neanderthal fell in the middle, with 26 differences. What's more, it was about
equidistant from all the modern populations to which it was compared,
suggesting that the modern and Neanderthal lineages had been evolving
independently for a long time.

Once the way had been shown, various research groups rapidly obtained
much more complete mtDNA genomes from more than a dozen Neanderthals found all over Europe. The result was always basically the same, and it
was additionally determined that both Neanderthals and modern humans
varied in their mtDNA sequences much less than chimpanzees do, suggesting that both hominid lineages had experienced relatively recent genomic
"bottlenecks." Evidently chimpanzees—partly because of longer speciation
times, but quite possibly also because they occupied less unsettled environments—have been unimpededly accumulating mutations in their mtDNA
for much longer than either *Homo neanderthalensis* or *H. sapiens* has.

Then the plot thickened. Using new high-throughput DNA sequencing
methods that mimic the passage of time by shredding the DNA into small
pieces before reassembling it in a computer, the Pääbo group contrived in
2010 to extract most of a Neanderthal nuclear genome from fragments of
bone unearthed at the Croatian cave of Vindija. In 2013 they followed this
triumph with an even more comprehensive nuclear genome from a fossil
found in the Altai Mountains of southern Siberia. These were staggering
technical accomplishments: in contrast to the merely 16,500-nucleotide-
long human mitochondrial genome, the human nuclear genome is some
3 billion bases long. Even worse, the mixing up of the nuclear DNA that
takes place between generations also means that it is hugely more complex
to analyze than its mitochondrial counterpart. The end result is vast data
sets that have to be crunched by some very heavy-duty computer algorithms
before the results are comprehensible to a human being. These algorithms,
very similar in principle to the ones that had been developed for use in nu-
merical cladistics, are necessarily built on a variety of assumptions about
evolutionary process that may be reasonable, but are not necessarily directly demonstrable. Still, changes in the DNA encode a vast amount of
population history, as well as the assembly instructions for each individual, allowing the group to demonstrate along the way not only that at least
some Neanderthals had possessed alleles suggesting red hair and light skin

color, but that they also had the human variant of FOXP2, one gene (among many) whose normal function is necessary to assure deficit-free speech.

The most unsurprising finding from the nuclear DNA work was that the Neanderthal and modern genomes are extraordinarily similar. Such resemblances are very hard to quantify exactly, but one method of comparing them suggests that we have 99.7 percent of our protein-coding genome in common with the Neanderthals. This compares with 98.8 percent similarity between humans and chimpanzees, which are also—in the greater scheme of things—very close relatives. The "molecular clock" initially suggested that the Neanderthal and modern lineages had parted ways at some time between about 300,000 and 700,000 years ago, a split time that has since fluctuated but, at the longer end, is compatible with the fossil record.

The researchers' most surprising conclusion was that some specific genomic similarities existed between Neanderthals and human populations outside Africa, the continent in which *Homo sapiens* evolved. The Pääbo team believes that this is due to hybridization in Europe between the resident Neanderthals and the descendants of some of the first modern *H. sapiens* to exit Africa, though any such interchange must have occurred very early on, since resemblances to Neanderthals are no greater among Europeans than they are among eastern Asians. The consequent gene flow into modern humans was estimated to have been on the order of 2 to 4 percent, although this number was subsequently reduced to 1.5 to 2.1 percent. Still, some scientists argue that these similarities might equally have been due to multiple modern human expansions out of Africa, some of which happened to conserve a signal derived from the ancient common ancestor with Neanderthals, while some didn't. The issue is still actively debated, but any reservations about what might actually have gone on as *H. sapiens* flooded into Europe did nothing to inhibit commercial laboratories from promising to tell anyone who coughed up a hundred bucks what percentage of "Neanderthal genes" they carry.

From the paleoanthropological point of view, however, what seems most important in the midst of all this is to remember that closely related species (like *Homo neanderthalensis* and *H. sapiens*) are always very leaky vessels, and a minor level of intermixing would hardly indicate that the Neanderthal and modern lineages were not historically individuated. There can still be little doubt that the two kinds of hominid had long since gone their separate evolutionary ways. Indeed, recent findings by the Pääbo group suggest that isolating mechanisms of the kind that Ernst Mayr saw as critical to complete speciation were actively evolving, for they found evidence among modern humans of "selection to remove genetic material derived from Neanderthals." To add to the complexity of the emerging picture, mathematical modeling strongly suggests that even the suggested 2 to 4 percent Neanderthal signal among modern humans could also have been achieved not only by demographic dynamics, but by a remarkably few instances of early mating.

Of course, to a morphologist the most telling thing of all is that the fossil record gives us precious little reason to suspect that any biologically meaningful melding occurred between the two distinctive kinds of hominid. *Homo neanderthalensis* maintained its morphological identity until it disappeared, and *H. sapiens* is, well, *nosce te ipsum*, as Linnaeus succinctly put it. That there may have been a little Pleistocene hanky-panky as the earliest *H. sapiens* swarmed into Europe is hardly surprising; we might even have acquired the odd useful gene from the Neanderthals. But the bottom line is that in meaningful systematic terms, the divorce had become final long before.

We have heard a lot in the press lately about epigenetics, a term normally used to describe heritable changes that do not result from changes in the actual base sequence of the DNA (i.e., in the genes). The commonest way of changing the function of DNA without changing its sequence is through a process called methylation, in which a tiny hydrocarbon is attached to one type of DNA base, in this case a C. One intriguing recent finding made possible by the sequencing of the Neanderthal genome is that at least some of the striking morphological differences between Neanderthals and modern humans may have epigenetic origins. These would have involved the alteration, by C methylation, of noncoding transcription factors belonging to what is called the "HOXD cluster." The regulatory genes of the HOXD cluster govern limb development by controlling the activity of the protein-coding genes involved in embryonic body plans. Human beings are notable for being much more slightly built than Neanderthals, a difference that appears to have been both recently and abruptly achieved. And at last, a corner of the curtain is being lifted on what exactly was going on at the genomic level to allow the two diverging lineages to follow different developmental pathways. Still, it will be necessary to know a lot more than we do now before we can properly understand all the implications of the Pääbo group's suggestion that, since the divergence, the genes involved in skeletal morphology have changed a lot more in the Neanderthal lineage than in the modern human one. To a morphologist this conclusion appears odd, because a robust and basically Neanderthal-like condition seems to have been standard for earlier members of *Homo*, while the gracile *Homo sapiens* appears uniquely evolved.

Another huge surprise came from the Pääbo group in 2010 and 2013, when mitochondrial and nuclear genomes, respectively, were obtained from an otherwise unidentifiable hominid finger bone recovered at the Denisova cave in southern Siberia. The owner of this bone was thought to have lived between about 30,000 and 50,000 years ago, a time when Neanderthals are known also to have been in the region. What's more, that is also a period during which early *Homo sapiens* might have been expected to pass through the area as its population expanded toward the east. Genomic analysis showed that this morphologically inscrutable bone had belonged to a female who, though related to Neanderthals, had an identifiably distinct DNA signature in both her mitochondrial and nuclear DNA. The researchers named the population to which she had belonged the Denisovans, and found that its members had shared a more recent common ancestry with Neanderthals than with modern humans. DNA from a rather strange isolated tooth found at Denisova told a similar story; and although the investigators declined to declare that the Denisovans belonged to a species distinct from Neanderthals, their discussion strongly suggests they believed this to be the case. But that wasn't all. They also concluded that this population, which they thought had probably been widespread across Asia at some point during the Pleistocene, had exchanged genes—at a low level—with modern humans as the latter emerged from Africa and spread across Asia. It even seemed to have intermixed—again, at a low frequency—with another, and otherwise unknown, population of hominids that was neither modern nor Neanderthal. Interestingly, Denisovan genes turned out to be represented disproportionately among modern Melanesians, suggesting a somewhat higher frequency of interbreeding as the ancestral Melanesians crossed Asia on the way to their ultimate insular destination.

The emerging story of what happened as early *Homo sapiens* spread out of Africa and encountered resident hominids in other places is thus remarkably complex, even though we currently have only a handful of ancient genomes. But what the genomic data seem to be telling us right now is that episodes of low-frequency admixture with resident hominid populations were a rather routine occurrence as our precursors rapidly fanned out across the habitable Old World. Since all participants in this process were very closely related, this is hardly unexpected. But the results of this intermixing—however it might have been achieved—were biologically trivial (though it has recently been argued that an allele found in Tibetans that helps them cope with the high altitudes has Denisovan origins). This is intriguing; but beyond the acquisition of the odd gene or two, in retrospect any intermixing does not seem to have affected the future histories of any of the entities involved. All of the actors in this drama that have yet been identified seem to have become sufficiently differentiated before intermixing to have been on their own independent evolutionary trajectories. In support of this conclusion, lingering archaic genetic elements seem to be scattered throughout the DNA of modern humans, rather than concentrated in those areas of the genome that one might expect to contribute notably to specific and functionally meaningful aspects of development.

Apart from the Neanderthals and Denisovans, the only extinct hominid yet to have had its DNA successfully sequenced is a fossil from the Sima de los Huesos, at Spain's Atapuerca. As you'll recall, the morphology of the Sima hominids clearly suggests that they were antecedent to the Neanderthals, or at least had belonged to a primitive sister species. An almost complete mtDNA sequence supports this assignment, showing the Sima population to be related to both Neanderthals and Denisovans, though closer to the latter. This close relationship makes one wonder how many fossils that we identify morphologically as Neanderthals might themselves actually be more closely related to Denisovans. Only time and the inexorable advance of technology will clarify this.

What the genomic data most eloquently emphasize is just how closely related all the variants of *Homo* are, as a result of the remarkable speed of evolutionary events in our genus during the Pleistocene. What is more, no other group in the entire living world is as closely scrutinized by scientists as ours, ensuring that the most inconsequential differences are agonized over. These two factors combine to make it starkly evident just how difficult it is in practice to recognize clear boundaries in Nature. If there is one thing on which my colleague Milford Wolpoff and I can agree, it is that we are dealing here with a very close-knit group in which lines are inherently going to be difficult to draw. But the lesson the genomic evidence has taught us so far, despite the blurriness of those lines, is that distinctive hominid lineages have both existed over time, and contrived to maintain their historical identities.

HOMO NEANDERTHALENSIS: PORTRAIT OF A SPECIES

During the final decade of the last century and the first one of this, a steady stream of new discoveries came from Europe, both of Neanderthal fossils and of more ancient hominids. The most important of those earlier finds came from a number of ancient cave-fill sites in that extraordinary Atapuerca region of northern Spain. The oldest of them was from a site known as the Sima del Elefante, and consisted of the front bit of a toothless jaw. It was not possible to say much from this fragment other than that hominids were already around at Atapuerca by about 1.3 million years ago. This was already significant, pushing back the earliest definitive hominid occupation of Europe by a couple of hundred thousand years and giving the crude stone tools from the Sima del Elefante a putative maker. But yet more intriguing was a suite of fragmentary hominid fossils, some 800,000 years old, from a nearby stone tool locality called the Gran Dolina. Paleoanthropologists cannot agree on whether or not the Gran Dolina hominids, which received the name of *Homo antecessor* in 1997, are more closely related to *H. heidelbergensis* or to the Neanderthals, or equally to both. Part of the problem may be that most of the bones are immature, but the new species has nonetheless already become a convenient umbrella for a miscellaneous grouping of early hominids from Europe, including the Sima del Elefante form and the maker of some possibly 900,000-year-old footprints discovered not long ago on a beach at Happisburgh (pronounced "Hazeboro") in eastern England.

What many find most intriguing about the Gran Dolina *Homo antecessor* is that the fossils bear clear signs of cannibalism. Found just inside an ancient cave entrance where early hominids made tools and prepared their meals, the hominid bones are broken in exactly the same manner as the mammal bones occurring alongside them. Further, they show cut marks and fracturing and pitting of the kind normally associated with butchery. Their finders' consequent argument that these unfortunate individuals were feasted on by their fellows (or by some kind of hominid, at least) seems to be well founded. What is more, there is no evidence that the context in which all this happened was a ritual one, so cannibalism evidently did not mean to *H. antecessor* what it has typically meant to modern human anthropophages. Indeed, because these hominids lived in a fairly rich habitat, and the practice seems to have been sustained over time, the Gran Dolina researchers suggest that what seems a repugnant practice to us was actually a routine part of *H. antecessor*'s subsistence strategy.

Interestingly, although no traces of cannibalism have been found on the extensive bone collection from the nearby Sima de los Huesos, something similar may also have been true of these hominids' later Neanderthal relatives. As we've seen, the Neanderthals flourished (at least in low densities: the latest molecular work, as well as the sparseness of their living sites, suggests they were always pretty thin on the ground) in Europe and western Asia (as far east as Siberia's Altai Mountains) between around 40,000 and 200,000 years ago. Many Neanderthal bones have now been shown to bear the marks of deliberate defleshing; and at yet another northern Spanish site, evidence of a particularly poignant vignette of Neanderthal life has been uncovered. At the cave of El Sidrón, an entire 12-strong social group of Neanderthals is believed to have been massacred some 50,000 years ago in a single event. After lying around for an undetermined period, the bones of these unfortunate individuals eventually collapsed, through an earth fall, into the underground chamber from which they were excavated.

The bones of six adults, three adolescents, two juveniles, and an infant had been broken up and butchered, and lay among numerous stone cores and flakes that they may have been making or using at the moment when their group was surprised by its killers. The teeth of all the El Sidrón individuals showed stress defects of the kind that develop when food is short, and the researchers surmise that—unlike those *Homo antecessor* at the Gran Dolina—they were the victims of "survival cannibalism" on the part of another group of *H. neanderthalensis*. The remains preserved enough mtDNA to show that genomic diversity was low in the El Sidrón group, as seems to have been typical for Neanderthals. Still, all three adult males belonged to the same mtDNA maternal lineage, while the adult females belonged to different ones, suggesting that males formed the core of the social group while the females "married" out. And the small total number of individuals—12—seemed pretty much in line with other estimates of Neanderthal group size based on the extents of their living sites.

Even as sites such as El Sidrón were yielding unprecedented new perspectives on Neanderthal social life, we were also learning more about the physical distinctiveness of *Homo neanderthalensis*. The species was already known from many more individual fossils than any other extinct hominid, including several partial skeletons, and the well-documented robustness of its bones had already given the species a bruiser's reputation. But still, nobody knew exactly what kind of physical impression an adult Neanderthal would have made in life. The only way to determine this for sure was to assemble an entire Neanderthal skeleton, and that could only be done with elements mustered from several different sites. This daunting task was eventually taken on by my colleagues Gary Sawyer and Blaine Maley, working in my lab at the American Museum of Natural History. Using casts of bony elements from half a dozen different sites in almost as many countries—and, critically, with top-to-bottom continuity furnished by bits and pieces of a key skeleton from La Ferrassie in France—Sawyer and Maley pieced together an entire bony Neanderthal. When they first unveiled the finished product in 2001, it blew me away. I thought I knew the elements of the Neanderthal skeleton well, and of course in a sense I did; but seeing the whole individual standing before me was another kind of experience entirely. For the first time, I felt I had actually met a Neanderthal. And I saw immediately that it was a very different creature from me.

One critical feature was the torso. The pelvis, I already knew, would be much broader and more flaring than mine, and the limb bones would be clunkier, with bigger joints. But the Neanderthal's thorax was unexpectedly different. Gary Sawyer had labored mightily, trying to get his Neanderthal's rib cage to look like the barrel-shaped human model that everyone—including me—had illustrated in their textbooks; but in the end he had given up and had just followed the form dictated by the available fossil elements. As a result, his Neanderthal boasted a conical thorax that tapered upward from the broad pelvis to a narrow top, giving it an incredibly distinctive look. Mentally comparing its features with those of various other fossils, I realized two things. The first was that its bodily structure made the Neanderthal fairly representative of middle Pleistocene *Homo*. And the second was that our species, *Homo sapiens,* is amazingly derived. Our slender bodies and barrel chests (let alone our strange globular skulls, with their tiny faces tucked beneath the front of the braincase) are totally unlike anything ever seen on Earth before. We are, literally, without close precedent in the fossil record. And that is why it was so peculiarly gratifying to read the very recent molecular analysis that fingered large methylation changes in the HOXD cluster as a potential cause of the anatomical differences between "archaic" and "present-day" humans. Here is a gap in the fossil record that may really be telling us something about both evolutionary pattern and mechanism.

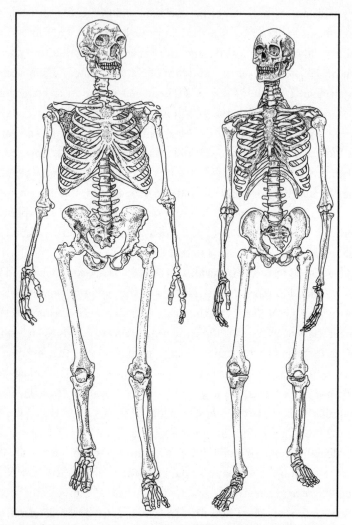

Comparison between a composite Neanderthal skeleton (left) and that of a modern human of similar stature. Note the numerous differences throughout the body, and especially in pelvic and thorax anatomy. In comparison to those of Homo sapiens the long bones of the Neanderthal are notably thick-walled, and the joint surfaces large. Neanderthal skeleton reconstructed by Gary Sawyer and Blaine Maley, art by Jennifer Steffey, based on a photograph by Ken Mowbray.

At the same time, of course, there is no doubt that for all their differences from us, the Neanderthals were a formidable presence on the planet.
As we come to know more about them, there is less and less doubt that
they deserve their own distinct identity—though when I have the temerity
to say in public that they are not *Homo sapiens*, I still sometimes find myself excoriated as a Neanderthal basher. Because the Neanderthals had big
brains (on average some 10 percent bigger than ours today, although comparable to those of the *H. sapiens* who lived alongside them because modern brains seem to have shrunk a bit), to many it somehow seems rather
shamefully discriminatory to banish them from our species. But I am sure
the Neanderthals never had any problem considering themselves intrinsi-
cally different from the oddly and unpredictably behaving hominids who
intruded into their territory some 40,000 years ago. For not only were the
Neanderthals anatomically distinctive, they were cognitively and behaviorally different from us as well. It is especially important to understand those
differences because there can be no doubt that *H. neanderthalensis* is the
best mirror we have in which to see the reflection of our own uniqueness.

The Neanderthals were excellent craftsmen in stone. They had mastered
the sophisticated art of making compound tools, with stone tips hafted into
handles. There is little question that they used stone-tipped projectile weapons. They were hugely effective hunters, and stable isotope methods (which
examine the ratios in their tissues of different isotopes of nitrogen) have
confirmed that, at least at some times and in some places, they hunted fearsomely big game such as woolly rhinoceroses and mammoths. Yet despite
this ability to specialize, Neanderthals remained flexible in their behaviors,
and as conditions changed they altered their diets. In the words of my colleague Ofer Bar-Yosef, they ate "what is there." In the northern latitudes at
which the Neanderthals typically lived, the obvious Ice Age dietary resource
was animal protein and fats, in the form of the big-bodied grazers teeming
on the tundra; but at the 40,000-year-old Belgian site of Spy, microfossils
recovered from calculus adhering to Neanderthal teeth indicate their owner
had also consumed a wide variety of plant foods, at least some of which had
been cooked. At a site in Italy, we see the same pragmatism reflected in hunting strategies that, as judged from bone refuse, differed between warmer and
cooler times. From studies like these, the Neanderthals emerge as adaptable
and resourceful—although, tellingly, their stone-working techniques didn't
vary much over the large extent of time and space they inhabited. Evidently

they followed the ancient hominid pattern of adapting old tools to new uses as environmental circumstances changed, rather than adopting the modern human expedient of inventing new technologies.

Most tellingly of all, though, the Neanderthals do not seem to have dealt with information exactly in the way we do today. What makes modern humans apparently unique in this regard is our ability to deconstruct our interior and exterior experiences into a vocabulary of mental symbols. We can then mentally combine and recombine those symbols according to rules, not just to describe the world as it is, but to reformulate it as it *might* be. This ability to manipulate abstract symbols to envisage a virtual infinity of possibilities is what makes our vaunted creativity possible; and despite some possible straws in the wind, any reasonable reading of the material record makes it evident that our particular form of inventiveness is unique to us. Yes, the Neanderthals do seem to have possessed empathy in some form. We may see this expressed, for example, in the burial of the dead, a practice they clearly indulged in from time to time, though typically very simply. Indeed, by promoting long-term preservation, burial almost certainly explains why we have so many Neanderthal fossils available to study. But Neanderthals only very occasionally—and then very arguably—placed anything that can convincingly be seen as symbolic goods in the graves to accompany the deceased. This contrasts strongly with the practices rapidly developed by the modern humans who entered their territory some 40,000 years ago.

All in all, the lives of the Neanderthals seem to have been distinguished by a *lack* of symbolic objects. Virtually all putative early symbolic pieces that are not at least reasonably associated with *Homo sapiens* have been challenged as such, by virtue of doubts either about their symbolic nature or about their association with Neanderthals—or any other archaic hominid. Most tellingly of all, what we can very confidently say is that symbolic objects did not play a routine role in Neanderthal lives. Something similar also applies to what we can reasonably infer about Neanderthal daily behaviors. The Neanderthal attitude toward cannibalism, for example, seems to have been disturbingly (to us) utilitarian, without any hint of the ritual that has invariably marked this rare behavior as somehow special when carried out by modern humans. In sum, what the Neanderthal record shows most eloquently is that it is possible to be very smart indeed without displaying either the particular kind of intelligence that modern human beings have, or the odd behaviors that result from it.

THE EMERGENCE OF US

The first decade of this century produced a bonanza of new information for anyone interested, as I increasingly was, in the origin of our species and its unusual proclivities. In 1998 I had published a book, *Becoming Human*, in which I sketched the outlines of what I thought was the most plausible scenario to account for the rapid and mind-boggling transformation of a rather unconventional ape into a creature that was entirely unprecedented in the history of the world. What we learned in the decade and a half that followed the book's publication hugely expanded the information that could be brought to bear on this issue, and it gratifyingly supported my earlier impressions.

A hugely important factor was the 2005 re-dating of Richard Leakey's basically modern Omo Kibish skull from southern Ethiopia to a whopping 195,000 years ago. This made the Omo specimen the most ancient *Homo sapiens* known; together with its 160,000 year-old compatriot from Herto, it established that anatomically modern humans were already around in Ethiopia not long after 200,000 years ago. From a strictly historical point of view this arrival was a momentous event, and ultimately it was going to be a fateful one for the entire world. But bearing in mind how biological and behavioral change had been out of phase throughout hominid evolution, in the short term you might not have expected it to be accompanied by any discernible behavioral change. And it wasn't. The stone tools deriving from the same sediments as the Omo cranium have been labeled "nondescript"; those at Herto were positively archaic, including as they did the very latest handaxes known from anywhere in Africa. As far as can be told, the appearance on Earth of *H. sapiens* was not accompanied by any behavioral fanfare (except perhaps that curious polish on the Herto infant skull). The first anatomically modern humans evidently continued to behave very much as their ancestors had.

But then there was that curious anatomy, possibly underwritten at least in part by that C-methylation of the HOXD cluster. Whatever the exact genomic alteration was, it must have arisen spontaneously in a precursor population that was small enough for the new configuration to have become fixed rapidly—in precisely the kind of process that geneticists have tended to neglect, as a result of their preoccupation with long-term selection pressures and the like. And it clearly had cascading effects throughout the skeleton, from the balloon-like head of the new form to its delicate feet. What's more, there is no compelling reason to conclude that those effects should have been limited to the structure of the skeleton. Because the genome is so frenetically busy, the very same regulatory change may well have had soft-tissue ramifications that extended to aspects of brain structure.

Exactly what the key neural modification was that permitted humans eventually to acquire symbolic cognition remains anyone's guess, although it is a good bet that it did not involve anything specifically visible in the imprints the brain makes on the inside of the skull: the paleoneurologists who study such things disagree about what they are seeing in the fossils with typical paleoanthropological vigor. Neuroanatomists are currently paying a lot of attention to the microarchitecture (cellular structure) of the neocortex, the thin but complex outer layer of the brain, as a key factor. However, most suggestions about what it is in the brain that makes symbolic thought possible still involve some form of enhancement of the abilities of separate specialized parts of the neocortex to "talk" to each other, and to make complex associations among their inputs and outputs.

One of the most intriguing recent molecular findings is that, compared to apes, metabolic activity in the human prefrontal cortex has accelerated hugely in the human lineage, and certainly this area of the neocortex figures hugely in anyone's assessment of what makes the modern human brain special. Still, regardless of exactly what the magic neural ingredient consisted of in structural terms, it seems altogether reasonable to believe that it was acquired as an integral part of the larger developmental reorganization that gave rise to the distinctive modern body form. An early appearance of this kind would fulfill the requirement that any new feature must already be in place before you can use it; indeed, it seems evident that all heritable novelties must arise as exaptations (existing features that are co-opted for future uses) rather than expressly as adaptations to current conditions. This is certainly what happened when birds learned how to use the feathers they already had to fly, or when the ancestors of land vertebrates acquired rudimentary limbs while they were still aquatic.

In any event, the Omo and Herto fossils tell us that anatomically modern humans had been around for quite some time before one population, again most likely a small isolate that had been battered and shrunken by changing environments, began to behave in a new way. Or maybe, since the potential was already there in all anatomically modern *Homo sapiens,* faltering starts were made in more than one part of Africa; we just won't know until the record is a lot better than it is now. What we do know, though, is that at around 100,000 years ago straws begin to appear in the wind. And straws are almost literally what they are. Inevitably, the archaeological record is never going to be easy to read in respect to ancient humans' cognitive styles, because mental processes never leave direct material traces. The best we can do here is to try to find indirect archaeological proxies for the kinds of behavior we associate with modern people .

When we do this, it is important to remember that looking for evidence of *symbolic* behaviors is entirely different from looking for evidence of *complex* behaviors, because the immediate precursors of *Homo sapiens* also indulged in some extremely complex activities, such as making and hafting sophisticated stone tools. As I have already emphasized in the case of the Neanderthals, you can evidently be highly intelligent in an intuitive way without processing information in the unique symbolic human manner. Given that evolution always has to build on what was there before, it seems reasonable to conclude that modern humans achieved their cognitive uniqueness by grafting the symbolic faculty onto a preexisting high intelligence of the ancestral intuitive kind—possibly exemplified by that of *H. neanderthalensis*—rather than by replacing that older style of intelligence wholesale. And it seems reasonable to conclude, if this is the case, that the only reliable proxies for symbolic cognition will be explicitly symbolic objects, followed by less direct evidence for symbolic behaviors.

But just what is a symbolic object? Opinions have varied on this one. The trickiest and most relevant cases here are the grinding of ochre and the piercing of small marine shells, presumably to permit stringing. The mere presence of ground ochre at a site is arguable as a proxy for symbolic behaviors because, though people have often used ochre in bodily decoration, the substance also has significant practical uses. Ochre was ground quite frequently in the European Upper Paleolithic and has occasionally been reported from Neanderthal sites, though its context is uncertain. What is certain, though, is that over the same period the practice became much more common over the course of the Middle Stone Age (MSA) of Africa. Ochre use has a long history in Africa, going back almost to the beginnings of the MSA over a quarter of a million years ago. The consensus is that the initial uses of ground ochre in the MSA were strictly utilitarian, but that by the time ground ochre was used around 100,000 years ago to color the putatively decorative pierced shell beads found at sites in northern Africa and the nearby Levant, things had changed. In every modern society in which it has been documented, bodily decoration (by direct painting or by adornment with jewelry) has had explicitly symbolic connotations: of wealth, taste, social status, clique membership, age cohort, and so forth. And it seems entirely legitimate to believe that by the later stages of the MSA, ochre paints had already come to take on a symbolic significance.

This strong suggestion of a radically new element in hominid life in Africa following about 100,000 years ago is powerfully supported by other findings in the same continent. In 2002 Christopher Henshilwood and colleagues reported a remarkable finding at Blombos Cave, on South Africa's southern coast: a couple of smoothed ochre plaques into which a repetitive geometric motif had been engraved. At some 77,000 years old, these are the most ancient overtly symbolic pieces yet described. What the design was intended to convey will probably never be known, but the two pieces were not of exactly the same age, suggesting that this motif had retained its meaning over time. And over space: another piece of ochre, of around the same date but found some 250 miles distant, bears a not dissimilar design.

That a new spirit was abroad in southern Africa by 70,000 to 80,000 years ago was emphasized in 2009, when researchers digging caves at Pinnacle Point, not far from Blombos, announced that the people who had lived there 75,000 years ago had used an extremely complex technology to harden an indifferent toolmaking material called silcrete into an excellent one. This unprecedented technique, which involved heating and cooling the

One of the smoothed and engraved ochre plaques from Blombos Cave, South Africa: the world's most ancient overtly symbolic objects. Drawn by, and courtesy of, Patricia Wynne.

silcrete in a series of exact steps, could almost certainly never have been invented or carried out by humans who lacked fully modern powers of reasoning and forward planning.

There is a lot of other evidence that the unique human way of viewing the world, and of manipulating information about it, was actively emerging in the African continent after about 100,000 years ago. Before that time, unequivocal evidence for symbolic behaviors is rare or lacking in the record; after it, such evidence rapidly accumulates. So, what could possibly have happened to spur members of the already-established species *Homo sapiens* to begin using their brains in this radically new way? After all, the transition from the nonsymbolic to the symbolic cognitive condition was, on the face of it, an extremely improbable one; it certainly could not have been predicted from anything that went before. Indeed, the only reason we have for believing such a transition *could* ever happen is that it so evidently *did* happen. To understand the geometry of these improbable events, we have first to remember that the neural potential for the new cognitive style was necessarily already in place (having almost certainly been there since the very birth of *H. sapiens* as a distinctive anatomical entity). The presence of the enabling biology was, after all, a prerequisite for anything to happen. And this in turn means the stimulus for the cognitive shift concerned must have been not a structural but a behavioral or, even more explicitly, a cultural one. And, since the transition was more or less instantaneous in evolutionary terms, we have to discard the most popular category of explanations for it, all of which in one way or another involve gradual pressures toward greater intelligence as the social milieu became more complex.

This leaves us with only one obvious candidate for this cultural stimulus: the invention of language. Like symbolic thought, language involves mentally creating symbols and reshuffling them according to rules; so close are these two things that it is virtually impossible for us today to imagine one in the absence of the other. What is more, it is relatively easy to envisage, at least in principle, how the spontaneous invention of language in some form could have started those symbols chasing around early modern human minds in a structured way. By the same token, it is no problem to understand how language and its cognitive correlates would rapidly have spread among members, and ultimately populations, of a species that was already biologically enabled for them. What is more, if language was invented by members of *Homo sapiens*—rather than being a deep property of the hominid clade, as many have thought—the peripheral vocal structures needed to express it in the form of speech were obviously already in place when it appeared, as they had been ever since the anatomical species had emerged. This geometry of events obviates any need for special explanation of how the necessary central and peripheral components of the language system had contrived to coordinate their arrivals.

It is possible to argue that, seen from the perspective of the Earth's biota, the acquisition of modern cognition was historically a less important event than the much later invention—and Faustian bargain—of agriculture and settled life. After all, the first cognitively modern humans retained the basic hunting-gathering lifestyle of their predecessors for many tens of thousands of years following this epic event, while it was the later radical economic change that placed humans in intellectual opposition to Nature, and started the human population on its inexorable path of increase. This having been said, the initial cognitive shift nonetheless made possible everything that was to come, and ushered in an era in which cultural and technological change became the norm, rather than the occasional exception. We currently glimpse the details of this extraordinary African revolution only very indistinctly, but knowing more about what exactly it involved will be essential to answering fundamental questions both about our essential nature and about our role in the world.

Nevertheless, while the emergence of *Homo sapiens* as a recognizable anatomical entity is still only very sketchily documented by fossils, it was clearly an event, and *not* a long-drawn-out process. Similarly, even though we have a very long way to go before we will have anything like a full appreciation of the complex behavioral events that unfolded during the later MSA, it is equally evident that the behavioral transformation this short period of human prehistory encapsulated was amazingly rapid. Although becoming fully human was thus discernibly a two-stage phenomenon, the whole thing was accomplished in an evolutionary eyeblink. That much seems well established. And if we perceive the abruptness of this fateful transition correctly, there is no way our unusual cognitive capacities can be the perfected products of long-term selective pressures. Thinking creatures we may be, but Nature has clearly not molded us for any specific purpose. What we are is up to us.

WHY DOES IT MATTER HOW WE EVOLVED?

WHAT YOU HAVE JUST READ IS A VERY PERSONAL ACCOUNT OF how paleoanthropology came by its received wisdoms. While there are a lot of very smart people out there who will disagree with much of what I've said, I doubt that many would take issue with everything. Most importantly, almost everyone will agree that the highly idiosyncratic history of paleoanthropology has deeply affected how we perceive our origins today. And this is important, because if the entire hominid fossil record were to be rediscovered tomorrow and analyzed by paleontologists with no horses already in the race, it is pretty certain that we would emerge with a picture of human evolution very different from the one we have inherited.

Still, an ever-growing fossil record has already made it inescapable that hominid history has not been the received saga of a lone hero battling from primitiveness to perfection over the eons, armed with nothing but natural selection and its own wits. Instead, our story has been a complex drama involving many players and an ever-changing environmental backdrop, along with numerous complex interactions and a substantial dollop of chance. And if this was the case, the conclusion must be that a great deal of the evolutionary change in our past was due at least as much to competition among populations and species in this unforgiving arena as it was to a reproductive race between individuals. The implications of the traditional and the emerging accounts of hominid evolution are very different; and the difference is much more than simply a formal matter of record, because the two views have profoundly dissimilar implications for the kind of creature human beings are today.

So let's take a moment to recap. We cannot understand how we came by our current assumptions about our past, and about the processes that governed it, without looking back at how our predecessors saw things. At the very beginning, Charles Darwin shied away from the fossil record, preferring to make a remarkable set of inferences about our evolutionary past from the meager evidence then available on our living primate relatives. This already put paleoanthropology on an unusual footing relative to the other paleontological sciences, but just as importantly it led Darwin to place less emphasis on those things that make us distinctive, and more on our continuities with the rest of Nature. In contrast, Darwin's contemporary and defender Thomas Henry Huxley not only disagreed strongly over whether evolution was necessarily the gradual, continuous process that Darwin conceived, but he was also much readier to confront the few actual hominid fossils available. Still, Huxley nonetheless contrived to find his own disingenuous arguments for cramming the gloriously distinctive Neanderthal fossil into our own species, *Homo sapiens*—thereby giving birth to paleoanthropology in a climate of hominid exceptionalism.

As hominid fossils slowly began to accumulate over the last few decades of the nineteenth century, paleoanthropology became the domain of human anatomists. A hallmark of this profession was an obsessive interest in the morphological variation that occurred within its sole subject species. When the anatomists turned their attention to fossils, this intense focus tended to express itself in two ways. Immediately, it showed in a conspicuous inattention to the significance of zoological names—which were, after all, the province of humble systematists. Later on, it was also expressed in a growing insensitivity to the significance—and systematic import—of the great morphological diversity that was revealed by an expanding human fossil record.

In the mid-twentieth century, under the traumatizing influence of Ernst Mayr, paleoanthropology's earlier disregard of evolutionary process yielded to worship of a "hardened" form of the New Evolutionary Synthesis. Species were now ephemeral segments of evolving lineages, within which evolutionary change equated directly with the slow accumulation of gene frequency changes under the control of natural selection. Paleoanthropologists may have so readily accepted this fundamentalist view not only because it filled a theoretical vacuum at the center of their profession, but also because it fit perfectly with the intuitively attractive idea of reconstructing human biological history by simply projecting today's single hominid species back into the deepest recesses of time. Mayr's effect on hominid systematics was equally profound, expressing itself in a hugely minimalist taxonomy. In a field littered with extraneous names, this was no bad thing in itself, but it placed paleoanthropology in an intellectual straitjacket. Instructively, when Mayr tried to do the same thing in his own discipline of ornithology, his colleagues reacted very differently, and soon began to develop species concepts that better reflected the actual bird diversity they perceived. In contrast, in paleoanthropology even tacit acknowledgment of past hominid diversity took decades, and the discovery of a huge variety of new fossils.

Still, once diversity began to be admitted, a new picture of hominid evolution began to develop rapidly—albeit in a totally *ad hoc* fashion due to the pressure of new discoveries, rather than because paleoanthropologists had any strong desire to bring their discipline more closely into line with other areas of paleontology. As a result, when I published the first in a series of tentative genealogical trees of the hominids in 1993 (see figures on pages 216 and 217), it featured only 12 hominid species spanning the period from 4 million years ago to the present. In stark contrast the latest version of this tree, published a mere 20 years later, contained twice as many species scattered over the last 7 million years. But both trees show the same two key things: that several hominid species had typically coexisted at any one point in time, and that *Homo sapiens* is the exception, rather than the rule, in being the lone hominid species on the planet. There is evidently something unprecedented about our species that makes it both intolerant of competition and uniquely able to eliminate it.

The scientists who have asked most explicitly what that something might have been are the evolutionary psychologists. Imbued with the hardened Synthesis—and taking as their canonical text Darwin's prediction in the *Origin of Species* that "in the distant future . . . psychology will be based on a new foundation, that of the necessary acquirement of each mental power and capacity by gradation"—evolutionary psychologists have

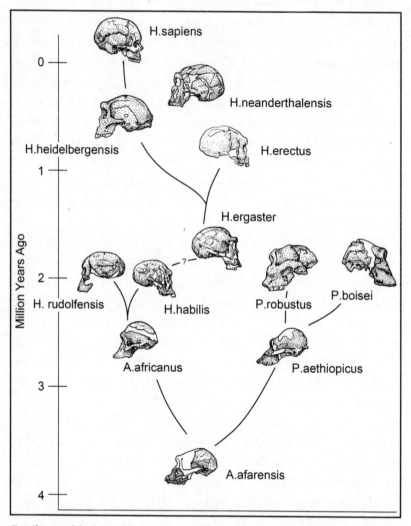

Family tree of the hominids, based on fossil materials known in 1993. See text. Drawn by Diana Salles.

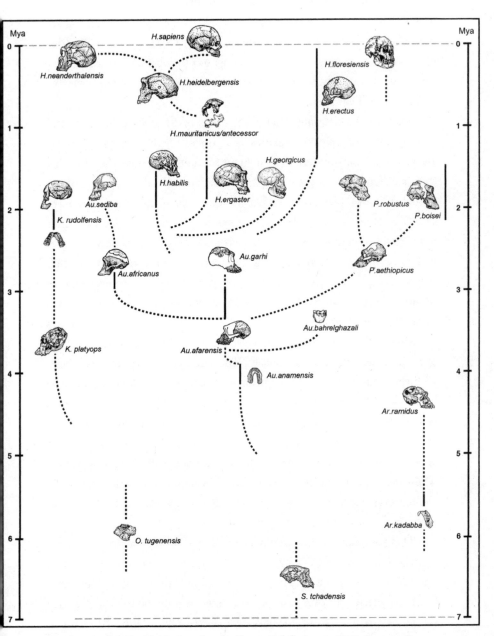

Family tree of the hominids, based on fossil materials known in 2012. See text. Drawn by Jennifer Steffey.

assumed that human behaviors have been fine-tuned by long-term natural selection, a process so gradual that selection has been unable to keep up with the vast, rapid, and largely self-created changes that have taken place in the human environment since the alarmingly recent adoption of settled life. By this reckoning, we human beings so often behave weirdly simply because slow natural selection has not yet had time to adjust our behaviors to our new milieu.

There can be little question that, by meticulously documenting our frequently unfathomable complexities, evolutionary psychologists have made an enormous contribution to our own understanding of ourselves. But what they have tended to forget is that most behaviors do not exist in isolation. Rather, each occupies a place on a continuum. Altruism—the rendering of service to others at significant cost to oneself—is a case in point. Certainly, individuals sometimes exhibit self-sacrificing behaviors; but let us never forget that hugely selfless deeds are invariably counterbalanced at the other end of the spectrum by acts of monstrous selfishness. Whatever kind of situation or behavior we may be talking about, most of us find ourselves somewhere in the middle most of the time, simply tending to help others out when the cost to ourselves is not too great—after all, we are members of an unusually cooperative species. The same pattern applies elsewhere, with the result that usually the behaviors exhibited by *Homo sapiens* as a whole—or by populations within it—are best described statistically, by a bell curve, and not by individual descriptors. Are individual humans selfish? Yes. Are they altruistic? Also yes. But obviously, neither designation applies to every individual, or even to any individual all of the time. Which is why the human condition over which philosophers and evolutionary psychologists spend so much time agonizing remains so elusive. In the human case one size never fits all, and never will.

This mundane reality would demand special explanation if our received origins myth were indeed accurate, and eons of natural selection had fine-tuned us to be the kind of creatures we are. After all, if our behaviors had been selected for over vast spans of time, we individuals, as well as our species as a whole, would inevitably be genetically predisposed to act in certain ways. But the reality is different. From the very beginning, far from having been a process of continuous refinement, hominid evolution clearly involved ongoing spontaneous experimentation as new species were regularly pumped out to compete on the ecological stage. Our species happened to be the one out of many that emerged triumphant from the resulting process of triage, and then employed its unprecedented and recently acquired cognitive capacity to see off all other pretenders to its niche.

This new capacity was based on a long series of neural acquisitions that had doubtless boosted the information-processing capabilities of the hominid brain over an extended period of time. But it was only suddenly—very recently, and in one sole lineage—that basically ancient ways of dealing with information were overlain—and, significantly, not replaced wholesale—by a radically new cognitive mode. This new mode was not only *exa*ptively rather than *ada*ptively acquired, but its properties were emergent, unpredicted by anything that had gone before. In other words, the novel way of processing information had not evolved *for* anything. It had just appeared, and the biology that made it possible evidently lay unexploited for some time, lying fallow until its possessor actively discovered its new uses.

This messy process explains our apparently contradictory cognitive condition. It explains why we have such brilliant rational abilities yet so often behave irrationally. It shows us why we so frequently use the unprecedented communicative potential of language to obfuscate and tell lies. It explains why we sometimes cannot justify our actions even to ourselves; why we are very poor judges of risk; why we can see environmental disaster approaching yet cannot bring ourselves to do anything to avert it; and why we reason so powerfully, yet make so many dreadful decisions. All this, and much more, occurs because the rational and the irrational constantly jostle in our heads, combining to make us the simultaneously creative yet reflexive creatures that we are. This extraordinarily paradoxical outcome would never have been predicted by gradual evolution under natural selection. Our gloriously untidy and creative condition could only have come about by an entirely adventitious series of events, though each one was built on—and constrained by—what existed before. So although our brains are without any question the products of a long and accretionary history that ultimately dates back hundreds of millions of years—and nothing would be the same today if any of the many events in that history had not occurred—we have evidently *not* been molded by Nature to any easily describable behavioral condition. We are not optimized for anything.

We can, of course, blame the historical and functional quirks of the evolutionary process for a lot—creaky backs, fallen arches, and gastric reflux come immediately to mind. But because our cognitive peculiarities were both recently and adventitiously acquired, we cannot blame evolution for how we individually behave. How we do that is ultimately up to each one of us individually, though we are also molded by our experience growing up in both family and society. Critically, the element of volition with which our untidy evolutionary background endows us is the fount not only of our free will, but also of our personal responsibility.

In contrast to the earlier perspective that saw the roots of Asians, Europeans, and Africans as extending deep into the evolutionary past, the extreme youth of our species also reminds us of the hugely epiphenomenal nature of the variations we see within it. Although our human eyes are keenly sensitive to physical differences among people of different geographic origins—in a way in which they may not be to the same kinds of variation in other species—in reality the differences we perceive among humans are of astonishingly recent origin. All have been acquired since *Homo sapiens* originated a mere 200,000 years ago. What's more, most of them have arisen in the 60,000 years or so since our early symbolic ancestors first left Africa. The large majority of those variations seem to be largely random with respect to the environment, although a few do seem to have adaptive significance, the most obvious among them being skin color. Here we find a clear trade-off: dark melanin pigment is essential for protecting delicate skin against the damaging tropical sun, but it impedes the vital synthesis of vitamin D in the weaker light of higher latitudes. However, even in this classic instance, recent genomic research indicates that the superficially uniform trait of dark skin color has actually been independently acquired in numerous different tropical populations, which have achieved it via distinct genomic pathways.

Particular markers in the genome have turned out to occur at different frequencies in peoples originating in different parts of the world, so it has lately become fashionable in some circles to revive the discarded notion that human "races" are biologically meaningful units. Thus, because using enough such markers allows a commercial laboratory to tell you, within a couple hundred miles, where one of your ancestors came from (or indeed, for a few bucks extra, what percentage of "Neanderthal genes" you possess), your genes are said to tie you to a geographic origin. And so, the argument continues, you belong to a meaningfully defined geographical group (your "race"). Yet the fallacy is painfully self-evident: you have many ancestors, and (perhaps most especially in America) it is highly unlikely that they all came from the same place.

Of course, it is undeniable that—on average—it is quite easy to tell an African from a European from an Asian from a native Australian. In general, people from each of those areas share features that were acquired by their ancestors as our species spread out over the Old World. Populations during that early period in our species' history would have been small and easily fragmented, and thus prone to acquire both genetic and cultural novelties. But especially since the initial adoption of sedentary lifeways at the end of the last Ice Age some 10,000 years ago, the human population has mushroomed, becoming both vastly denser and more individually mobile. Since that fateful point the biological story of *Homo sapiens* has been one not of differentiation but of reintegration, as populations have expanded and migrated on huge scales. The result has been a pervasive blurring of edges, making it impossible to draw any meaningful lines within our species.

The only substantial barriers to this ongoing process of biological integration are cultural ones. These exist because, while all human beings are born with the innate capacity to learn the language or the worldview of any society, our basic cultural attitudes become ingrained in our unconscious very soon thereafter. This is indeed problematic, because in later life different ways of mentally constructing the world can present even larger barriers to mutual understanding than those posed by linguistic differences. Nonetheless, even the unique human possession of complex culture can serve only to slow reintegration down. Under present demographic conditions, it will never quench it. Despite its huge geographic dispersion, *Homo sapiens* has contrived to remain one gloriously varied interbreeding population that—however reluctantly—remains rooted in the natural world.

This brings us back to the lemurs, with which this story began. *Homo sapiens* may indeed be unusual in some hugely significant respects, but as we learn more about it the history of our family comes increasingly to resemble that of the lemurs, at least in the overall pattern of diversity it displays. Those charismatic primates of Madagascar furnish us with an amazing example in the living world of how, through a vigorous process of evolutionary experimentation, a single group occupying a new environment can, and will, exploit every ecological possibility open to it—just as our own precursors clearly did. If we continue to regard ourselves as an exception to the rules of the evolutionary game as demonstrated by the lemurs, we will maintain a distorted perspective not only on our history, but on the creature that history has produced.

So, despite what we might have been taught, we are the pinnacle of nothing. Instead, we are simply one more twig on what was until very recently a luxuriant evolutionary tree. The recent drastic pruning of this tree—a product of our accidental uniqueness—has given us an entirely false view of our place in Nature. To keep our self-image in proper proportion, we should never forget that what succeeds in evolution is usually not optimization in the engineering sense, but simply whatever it is that happens to work in the current environmental marketplace. Special as we *Homo sapiens* like to think ourselves, no impartial observer would dispute that, after many millions of years of evolution, we are notably unperfected—and will almost certainly remain that way. This, above all, is why we should always remember that we are no exception to Nature's rules. Odd we may be; but we are nonetheless an odd *primate*.

NOTES AND BIBLIOGRAPHY

IN THIS SECTION I LIST, BY CHAPTER, ALL TITLES MENTIONED IN the text, and particularly all sources of quotations. For more comprehensive bibliographies on the areas concerned, please consult Tattersall (2009 and 2012) and Klein (2009). Detailed descriptions and black-and-white illustrations of almost all of the fossils mentioned in this book are to be found in the three volumes by Schwartz and Tattersall (2002, 2003, and 2005). Excellent color illustrations of many of the fossils featured may be found in Johanson and Edgar (2006), and elegant life reconstructions of some are presented by Gurche (2013).

Gurche, John. *Shaping Humanity: How Science, Art, and Imagination Help Us Understand Our Origins.* New Haven, CT: Yale University Press, 2013.

Johanson, Donald C., and B. Edgar. *From Lucy to Language.* 2nd ed. New York: Simon & Schuster, 2006.

Klein, Richard G. *The Human Career: Human Biological and Cultural Origins.* 3rd ed. Chicago: University of Chicago Press, 2009.

Schwartz, Jeffrey H., and Ian Tattersall. *The Human Fossil Record, Vol. 1: Terminology and Craniodental Morphology of Genus* Homo *(Europe).* New York: Wiley-Liss, 2002.

Schwartz, Jeffrey H., and Ian Tattersall. *The Human Fossil Record, Vol. 2: Craniodental Morphology of Genus* Homo *(Africa and Asia).* New York: Wiley-Liss, 2003.

Schwartz, Jeffrey H., and Ian Tattersall. *The Human Fossil Record, Vol. 4: Craniodental Morphology of Early Hominids (Genera* Australopithecus, Paranthropus, Orrorin*) and Overview.* New York: Wiley-Liss, 2005.

Tattersall, Ian. *The Fossil Trail: How We Know What We Think We Know About Human Evolution.* 2nd ed. New York: Oxford University Press, 2009.

Tattersall, Ian. *Masters of the Planet: The Search for Our Human Origins.* New York: Palgrave Macmillan, 2012.

CHAPTER 1: HUMANKIND'S PLACE IN NATURE

Samuel Johnson's first *Dictionary* was published in London in 1755. The definitive edition of Linnaeus' great work, *Systema Naturae*, was the tenth, of 1758. A readily available version of Aristotle's views on Nature is McKeon (2001). Lamarck's *Philosophie Zoologique* (1809) and Brocchi's *Conchologia Subalpennina* (1814) are these scientists' most comprehensive presentations of their thoughts on change; see El-dredge (2015) for these, as well as for the development of Charles Darwin's thought. Erasmus Darwin's musings on transformation appeared in his *Zoonomia* (1794). Uthman al-Jahith's ninth-century *Book of Animals* was quoted in the London *Daily Telegraph* by Jim al-Khalili on January 29, 2008. Charles Darwin's 1859 and 1871 works are widely available in reprint. The Wallace Sarawak paper is Wallace (1855) and the Ternate paper is Wallace (1858). Wallace's speculations on human evolution appeared in Wallace (1864).

Darwin, Charles. *On the Origin of Species by Means of Natural Selection. Or, the Pres-ervation of Favoured Races in the Struggle for Life.* London: John Murray, 1859.

Darwin, Charles. *The Decent of Man and Selection in Relation to Sex.* London: John Murray, 1871.

Darwin, Erasmus. *Zoonomia; Or, the Laws of Organic Life*, vol. 1. London: J. Johnson, 1794.

Eldredge, Niles. *Eternal Ephemera.* New York: Columbia University Press, 2015.

Johnson, Samuel. *Dictionary of the English Language.* London: Strahan, 1755.

Linnaeus, Carolus. *Systema Naturae.* 10th ed. Stockholm: Salvii, 1758.

Lyell, C. *Principles of Geology*, 3 vols. London: John Murray, 1830–33.

McKeon, Richard, ed. *The Basic Works of Aristotle.* New York: Modern Library Classics, 2001.

Wallace, Alfred R. On the law which has regulated the introduction of new species. *Ann. Mag. Nat. Hist.* 16 (1855): 184–95.

Wallace, Alfred R. On the tendency of species to depart indefinitely from the original type. *Proc. Linn. Soc. London (Zoology)* 3 (1858): 53–62.

Wallace, Alfred R. The origin of races and the antiquity of man deduced from the theory of natural selection. *Jour. Anthropol. Soc. London* 2 (1864): clviii–clxx.

CHAPTER 2: PEOPLE GET A FOSSIL RECORD

The essay "Some Fossil Remains of Man" is in Huxley (1863). See Schwartz (2006) for a discussion of Huxley's reasoning. For the English translation of Schaaffhausen's 1857 paper, see Busk (1864a). Quote is from Busk (1864b). For more detail on the Schaaffhausen/Mayer/Virchow controversy, see Trinkaus and Shipman (1993). *Homo neanderthalensis* was named by King (1863). For early archaeology, see Grayson (1983). Theunissen (1988) gives extensive detail on the career of Eugène Dubois and his relations with European colleagues such as Schwalbe. Mayr (1982) is an excellent source on early genetics. See Spencer (1990) for a balanced account of the Piltdown affair. Quote is from Gould (1978).

Busk, George. 1864a. On the crania of the most ancient races of man. *Nat. Hist. Rev.* 1:155–76.

Busk, George. 1864b. Pithecan priscoid man from Gibraltar. *The Reader*, July 23, 1864. Quoted in J. Reader, *Missing Links* (Boston: Little, Brown, 1981), 249.

Gould, Stephen J. Morton's ranking of races by cranial capacity. *Science* 200 (1978): 503–9.

Grayson, Donald K. *The Establishment of Human Antiquity.* New York: Academic Press, 1983.

Huxley, Thomas H. *Evidence as to Man's Place in Nature.* London: Williams and Norgate, 1863.

King, William. The Neanderthal skull. *Anthrop. Rev.* 1 (1863): 393–94.

Mayr, Ernst. *The Growth of Biological Thought: Diversity, Evolution, and Inheritance.* Cambridge, MA: Belknap Press, 1982.

Schwartz, Jeffrey H. Race and the odd history of human paleontology. *Anat. Rec. (New Anat.)* 289B (2006): 225–40.

Spencer, Frank. *Piltdown: A Scientific Forgery.* London: Natural History Museum/ Oxford University Press, 1990.

Theunissen, Bert. *Eugène Dubois and the Ape-Man from Java: The History of the First "Missing Link" and Its Discoverer.* Dordrecht/Boston: Kluwer Academic, 1988.

Trinkaus, Erik, and Pat Shipman. *The Neanderthals: Changing the Image of Mankind.* New York: Knopf, 1993.

CHAPTER 3: NEANDERTHALS AND MAN-APES

See Tattersall (2009) for details on continuing Neanderthal discoveries, Mauer, and glacial geology. The La Chapelle-aux-Saints skeleton was described and analyzed by Boule (1911–13); quote therefrom. For "The Neanderthal Phase of Man" see Hrdlička (1927); for the Broken Hill (Kabwe) cranium, see Woodward (1921). The initial description of *Australopithecus africanus* was by Dart (1925), from which quotes are taken. Black (1927 and 1931) contains first descriptions of *Sinanthropus pekeinensis* and of fire at Zhoukoudian, respectively. Weidenreich (1939) elaborated most comprehensively on evidence for cannibalism at Zhoukoudian. Boule's interpretation appeared in 1938. See Koenigswald and Weidenreich (1939) for synonymy of *Pithecanthropus* and *Sinanthropus,* and Weidenreich (1947) for his grid diagram. Mayr (1982) provides an account of the events leading up to the formulation of the New Evolutionary Synthesis—see also Eldredge (1985). The canonical texts of the Synthesis are Dobzhansky (1937), Mayr (1942), and Simpson (1944). The story of discoveries in South Africa is well told by Clark (1967), while Gregory and Hellman (1939) first formally placed the australopiths in Hominidae. Broom and Robinson noted diversity at the South African cave sites in 1949. Dart described the hominids and the osteodontokeratic culture from Makapansgat in 1948 and 1957, respectively. The Mount Carmel hominids were monographed (source of quotes) by McCown and Keith in 1939. Dobzhansky articulated his view of the human fossil record (as quoted) in 1944.

Black, Davidson. On a lower molar hominid tooth from the Chou Kou Tien deposit. *Palaeont. Sinica,* ser. D, 7 (1927): 1–29.

Black, Davidson. Evidences of the use of fire by *Sinanthropus. Bull. Geol. Soc. China* 11 (1931): 107–8.

Boule, Marcellin. L'homme fossile de La Chapelle-aux-Saints. *Annales de Paléontologie* 6 (1911): 1–64; 7 (1912): 65–208; 8 (1913): 209–79.

Boule, M. Le Sinanthrope. *L'Anthropologie* 47 (1937): 1–22.

Broom, Robert, and John T. Robinson. Man contemporaneous with Swartkrans ape-man. *Amer. Jour. Phys. Anthropol.* 8 (1949): 151–56.

Clark, Wilfrid E. Le Gros. *Man-Apes or Ape-Men? The Story of Discoveries in Africa.* New York: Holt, Rinehart and Winston, 1967.

Dart, Raymond A. *Australopithecus africanus:* the man-ape of South Africa. *Nature* 115 (1925): 195–99.

Dart, Raymond A. The Makapansgat proto-human *Australopithecus prometheus. Amer. Jour. Phys. Anthropol.* 6 (1948): 259–84.

Dart, Raymond A. The osteodontokeratic culture of *Australopithecus prometheus. Transvaal Mus. Memoir* 10 (1957): 1–105.

Dobzhansky, Theodosius. *Genetics and the Origin of Species.* New York: Columbia University Press, 1937.

Dobzhansky, Theodosius. On species and races of living and fossil man. *Amer. Jour. Phys. Anthropol.* 2 (1944): 251–65.

Eldredge, Niles. *Time Frames: The Rethinking of Darwinian Evolution and the Theory of Punctuated Equilibria.* New York: Simon & Schuster, 1985.

Gregory, William K., and Milo Hellman. The South African fossil man-apes and the origin of the human dentition. *Jour. Amer. Dental Assoc.* 26 (1939): 645.

Hrdlička, Aleš. The Neanderthal phase of man. *Jour. Roy. Anthropol. Inst.* 57 (1927): 249–74.

Koenigswald, Gustav. H. R. von, and Franz Weidenreich. The relationship between *Pithecanthropus* and *Sinanthropus. Nature* 144 (1939): 926–27.

Mayr, Ernst. *Systematics and the Origin of Species.* New York: Columbia University Press, 1942.

Mayr, Ernst. *The Growth of Biological Thought: Diversity, Evolution, and Inheritance.* Cambridge, MA: Belknap Press, 1982.

McCown, Theodore, and Arthur Keith. *The Stone Age of Mount Carmel,* vol. 2. Oxford: Clarendon Press, 1939.

Simpson, George G. *Tempo and Mode in Evolution.* New York: Columbia University Press, 1944.

Tattersall, Ian. *The Fossil Trail: How We Know What We Think We Know About Human Evolution.* 2nd ed. New York: Oxford University Press, 2009.

Weidenreich, Franz. Six lectures on *Sinanthropus pekinensis* and related problems. *Bull. Geol. Soc. China* 19 (1939): 1–110.

Weidenreich, Franz. Facts and speculations concerning the origin of *Homo sapiens. Amer. Anthropol.* 49 (1947): 187–203.

Woodward, Arthur S. A new cave man from Rhodesia, South Africa. *Nature* 108 (1921): 371–72.

CHAPTER 4: THE SYNTHESIS AND HANDY MAN

See Mayr (1950) for his view of the human fossil record (including quotes). Robinson (1953) finally persuaded Mayr of multiple hominid lineages. See Howell (1951) for Neanderthal analysis; Tuniz, Manzi, and Caramelli (2014) for a recent account of dating methods; and Movius et al. (1975–85) for Abri Pataud excavations. For first reports of Olduvai hominids, see L. Leakey (1959, 1961); for *Homo habilis,* see L. Leakey, Tobias, and Napier (1964); and for Natron fossil, see L. Leakey and M. Leakey (1964). L. Leakey, Evernden, and Curtis (1961) reported the first K/Ar date from Olduvai Bed I; M. Leakey (1966) reviewed the Oldowan culture; and Tobias and Koenigswald (1964) reported comparison of the Olduvai and Java fossils.

Howell, F. C. The place of Neanderthal Man in human evolution. *Amer. Jour. Phys. Anthropol.* 9 (1951): 379–416.

Leakey, Louis S. B. A new fossil skull from Olduvai. *Nature* 184 (1959): 491–93.

Leakey, Louis S. B. New finds at Olduvai Gorge. *Nature* 189 (1961): 649–50.

Leakey, Louis S. B., and M. D. Leakey. Recent discoveries of fossil hominids in Tanganyika: At Olduvai and near Lake Natron. *Nature* 202 (1964): 5–7.

Leakey, Louis S. B., J. F. Evernden, and G. H. Curtis. Age of Bed I, Olduvai Gorge, Tanganyika. *Nature* 191 (1961): 478–79.

Leakey, Louis S. B., P. V. Tobias, and J. R. Napier. A new species of the genus *Homo* from Olduvai Gorge. *Nature* 202 (1964): 7–9.

Leakey, Mary D. A review of the Oldowan culture from Olduvai Gorge, Tanzania. *Nature* 210 (1966): 462–66.

Mayr, Ernst. Taxonomic categories in fossil hominids. *Cold Spring Harbor Symp. Quant. Biol.* 15 (1950): 109–18.

Movius, Hallam L. Jr., and various collaborators. *Excavations at the Abri Pataud, Les Eyzies (Dordogne).* 4 vols. Cambridge, MA: Peabody Museum of Archaeology and Ethnology, Harvard University, 1975–85.

Robinson, John T. *Meganthropus,* Australopithecines and hominids. *Amer. Jour. Phys. Anthropol.* 11 (1953): 1–38.

Tobias, Phillip V., and Gustav H. R. von Koenigswald. Comparison between the Olduvai hominines and those of Java and some implications for hominid phylogeny. *Nature* 204 (1964): 515–18.

Tuniz, Claudio, Giorgio Manzi, and Davide Caramelli. *The Science of Human Origins.* Walnut Creek, CA: Left Coast Press, 2014.

CHAPTER 5: EVOLUTIONARY REFINEMENTS

Pilbeam and Simons (1965) reinterpreted the early human fossil record in the light of the Synthesis (all quotes from therein). Brace (1964) reviewed the Neanderthals. Simons first advocated *Ramapithecus* as a hominid in 1961. Landau (1984) emphasized narrative similarities between paleoanthropological scenarios and folktales. Goodman (1962) was an early advocate of applying an immunological approach to primate phylogeny. Sarich and Wilson (1966, 1967) advocated a late separation of humans and apes on immunological grounds. Quote is from Sarich (1971). Andrews and Tekkaya (1980) showed that *Ramapithecus* differed only slightly from *Sivapithecus,* which in turn had affinities with orangutans. Lipson and Pilbeam renounced *Ramapithecus* as a hominid in 1982. The "seed-eater" model was proposed by Jolly (1970). English translation of Hennig's book was published in 1966. The notion of punctuated equilibria was first published by Eldredge (1971), and under that name by Eldredge and Gould (1972).

Andrews, Peter, and Ibrahim Tekkaya. A revision of the Turkish Miocene hominoid *Sivapithecus meteai. Palaeontology* 23 (1980): 85–95.

Brace, Charles L. The fate of the "classic" Neanderthals: a consideration of hominid catastrophism. *Curr. Anthropol.* 5 (1964): 3–43.

Eldredge, Niles. The allopatric model and phylogeny in Paleozoic invertebrates. *Evolution* 25 (1971): 156–67.

Eldredge, Niles, and Stephen J. Gould. Punctuated equilibria: an alternative to phyletic gradualism. In *Models in Paleobiology,* edited by T. J. M. Schopf, 82–115. San Francisco: Freeman Cooper, 1972.

Goodman, Morris. Immunochemistry of the primates and primate evolution. *Ann. N. Y. Acad. Sci.* 102 (1962): 219–34.

Hennig, W. *Phylogenetic Systematics.* Urbana: University of Illinois Press, 1966.

Jolly, Clifford J. The seed-eaters: a new model of hominid differentiation based on a baboon analogy. *Man,* n.s., 5 (1970): 5–26.

Landau, Misia. Human evolution as narrative. *Amer. Scientist* 72 (1984): 262–68.

Lipson, Susan, and David Pilbeam. *Ramapithecus* and hominoid evolution. *Jour. Hum. Evol.* 11 (1982): 545–48.

Pilbeam, David R., and Elwyn. L. Simons. Some problems of hominid classification. *Amer. Scientist* 53 (1965): 237–59.

Sarich, Vincent M. A molecular approach to the question of human origins. In *Background for Man: Readings in Physical Anthropology*, edited by Phyllis Dolhinow and Vincent M. Sarich, 60–81. Boston: Little, Brown, 1971.

Sarich, Vincent M., and Allan C. Wilson. Quantitative immunochemistry and the evolution of primate albumins. *Science* 154 (1966): 1563–66.

Sarich, Vincent M., and Allan C. Wilson. Immunological time scale for hominid evolution. *Science* 158 (1967): 1200–03.

Simons, Elwyn L. The phyletic position of *Ramapithecus*. *Peabody Mus. Postilla* 57 (1961): 1–9.

CHAPTER 6: THE GILDED AGE

For the background to Louis Leakey's involvement in Ethiopia and Richard Leakey's in the Turkana Basin, see Morrell (1995). See Howell (1978) for his summary of the Omo hominids. *Paraustralopithecus aethiopicus* was named by Arambourg and Coppens (1968). Initial descriptions of early Turkana finds are in R. Leakey (1970, 1971, 1972, 1973, and 1974). Naming of *Homo ergaster* was by Groves and Mazak (1975). Evidence against single-species hypothesis was emphasized by R. Leakey and Walker (1976). Glynn Isaac's conclusions are summarized in Isaac (1978), whence quote. See Johanson and Edey (1981) and Kalb (2000) for very different accounts of Afar fieldwork. For nearby events and local color, consult Gitonga and Pickford (1995), as well as Morrell (1995). Early Hadar finds were described by the Hadar Group (1982); Bodo cranium by Conroy et al. (1978); Laetoli hominids by White (1977, 1980); and trackways in M. Leakey and Harris (1987). *Australopithecus afarensis* was named by Johanson, White, and Coppens (1978), and its unity and relationships were analyzed by Johanson and White (1979). Olson (1981) split the sample again. Lovejoy summarized his views on locomotion in *A. afarensis* in 1978, and developed his scenario on the causes of this unusual locomotor style in 1981. See review of competing arguments over early hominid locomotion by Harcourt-Smith (2014).

Arambourg, Camille, and Yves Coppens. Découverte d'un australopithecien nouveau dans les gisements de l'Omo, Ethiopie. *South Afr. Jour. Sci.* 63 (1968): 58–59.

Conroy, Glenn C., et al. Newly discovered fossil hominid skull from the Afar Depression, Ethiopia. *Nature* 275 (1978): 67–70.

Gitonga, Eustace, and Martin Pickford. *Richard Leakey: Master of Deceit*. Nairobi: White Elephant, 1995.

Groves, Colin P., and Vratislav Mazak. An approach to the taxonomy of the Hominidae: Gracile Villafranchian hominids of Africa. *Casopis pro Mineralogii Geologii* 20 (1975): 225–47.

Hadar Group. Special Issue: Pliocene Hominid Fossils from Hadar, Ethiopia. *Amer. Jour. Phys. Anthropol.* 57, no. 4 (1982): 373–724.

Harcourt-Smith, Will. The origins of bipedal locomotion. In *Handbook of Paleoanthropology*, vol. 3, 2nd ed. edited by W. Henke and I. Tattersall, 1919–60. Heidelberg: Springer, 2014.

Howell, Francis C. Hominidae. In *Evolution of African Mammals*, edited by V. J. Maglio and H. B. S. Cooke, 154–248. Cambridge, MA: Harvard University Press, 1978.

Isaac, Glynn L. The food-sharing behavior of proto-human hominids. *Scientific American* 238, no. 6 (1978): 90–108.

Johanson, Donald, and Maitland A. Edey. *Lucy: The Beginnings of Humankind.* New York: Simon & Schuster, 1981.

Johanson, Donald C., and Tim D. White. A systematic assessment of early African hominids. *Science* 203 (1979): 321–30.

Johanson, Donald C., Tim D. White, and Yves Coppens. A new species of the genus *Australopithecus* (Primates: Hominidae) from the Pliocene of eastern Africa. *Kirtlandia* 28 (1978): 1–14.

Kalb, Jon. *Adventures in the Bone Trade: The Race to Discover Human Ancestors in Ethiopia's Afar Depression.* New York: Copernicus, 2000.

Leakey, Mary D., and John M. Harris, eds. *Laetoli: A Pliocene Site in Northern Tanzania.* Oxford: Clarendon Press, 1987.

Leakey, Richard E. F. New hominid remains and early artefacts from northern Kenya. *Nature* 226 (1970): 226–28.

Leakey, Richard E. F. Further evidence of Lower Pleistocene hominids from East Rudolf, Kenya. *Nature* 231 (1971): 241–45.

Leakey, Richard E. F. Further evidence of Lower Pleistocene hominids from East Rudolf, North Kenya. *Nature* 237 (1972): 264–69.

Leakey, Richard E. F. Further evidence of Lower Pleistocene hominids from East Rudolf, North Kenya 1972. *Nature* 242 (1973): 170–73.

Leakey, Richard E. F. Further evidence of Lower Pleistocene hominids from East Rudolf, North Kenya 1973. *Nature* 248 (1974): 653–56.

Leakey, Richard E. F., and Alan C. Walker. *Australopithecus, Homo erectus* and the single species hypothesis. *Nature* 261 (1976): 572–74.

Lovejoy, C. Owen. The origin of man. *Science* 211 (1981): 341–50.

Lovejoy, C. Owen. Evolution of human walking. *Scientific American* 259, no. 5 (1988): 118–25.

Morrell, Virginia. *Ancestral Passions: The Leakey Family and the Quest for Humankind's Beginnings.* New York: Simon & Schuster, 1995.

Olson, Todd. Basicranial morphology of the extant hominoids and Pliocene hominids: the new material from the Hadar Formation, Ethiopia, and its significance in early human evolution and taxonomy. In *Aspects of Human Evolution,* edited by C. B. Stringer, 99–128. London: Taylor and Francis, 1981.

White, Tim D. New fossil hominids from Laetoli, Tanzania. *Amer. Jour. Phys. Anthropol.* 46 (1977): 197–230.

White, Tim D. Additional hominid specimens from Laetoli, Tanzania. *Amer. Jour. Phys. Anthropol.* 53 (1980): 487–504.

CHAPTER 7: MEANWHILE, BACK AT THE MUSEUM . . .

Eldredge and Tattersall (1975) and Tattersall and Eldredge (1977) discuss respectively the application of cladistics to paleoanthropology, and the complexity of phylogenetic hypotheses. Sangiran 17 was described by Sartono (1972), and early Javan dates were made by Swisher et al. (1994). Recent Ngandong age was determined by Swisher et al. (1996). Latest Zhoukoudian dates are from Shen et al. (2009). Chinese hominids are reviewed in Tattersall (2009); Petralona cranium by Stringer, Howell, and Melentis (1979). For Terra Amata, see de Lumley and Boone (1976); for the Arago hominids, see Schwartz and Tattersall (2002). For archaeology in this period, see Klein (2009). For the Atapuerca fossils, see Arsuaga et al. (1993, 2014). The Amud material was described by Suzuki and Takai (1970) and Rak et al. (1994), and the Kebara skeleton in Bar-Yosef and Vandermeersch (1991). For Jebel Qafzeh,

Jebel Irhoud, Saldanha, Border Cave, and Klasies River Mouth fossils, see Schwartz and Tattersall (2003).

Arsuaga, Juan-Luis, et al. Three new human skulls from the Sima de los Huesos Middle Pleistocene site in Sierra de Atapuerca, Spain. *Nature* 362 (1993): 534–36.

Arsuaga, Juan-Luis et al. Neandertal roots: Cranial and chronological evidence from Sima de los Huesos. *Science* 344 (2014): 1358–63.

Bar-Yosef, Ofer, and Bernard Vandermeersch. *Le squelette Moustérien de Kébara 2.* Paris: CNRS, 1991.

de Lumley, Henry, and Yves Boone. Les structures d'habitat au Paléolithique inféri-eur. In *La Préhistoire française*, vol. 1, edited by H. de Lumley, 635–43. Paris: CNRS, 1976.

Eldredge, Niles, and Ian Tattersall. Evolutionary models, phylogenetic reconstruc-tion, and another look at hominid phylogeny. In *Approaches to Primate Paleo-biology*, edited by F. S. Szalay, 218–42. Basel: Karger, 1975.

Klein, Richard G. *The Human Career: Human Biological and Cultural Origins.* 3rd ed. Chicago: University of Chicago Press, 2009.

Rak, Yoel, William. H. Kimbel, and Erella Hovers. A Neandertal infant from Amud Cave, Israel. *Jour. Hum. Evol.* 26 (1994): 313–24.

Sartono, Soedjojo. Discovery of another hominid skull at Sangiran, Central Java. *Curr. Anthropol.* 13 (1972): 124–26.

Schwartz, Jeffrey H., and Ian Tattersall. *The Human Fossil Record, Vol. 1: Terminol-ogy and Craniodental Morphology of Genus* Homo *(Europe).* New York: Wiley-Liss, 2002.

Schwartz, Jeffrey H., and Ian Tattersall. *The Human Fossil Record, Vol. 2: Cranio-dental Morphology of Genus* Homo *(Africa and Asia).* New York: Wiley-Liss, 2003.

Shen, Guo., et al. Age of Zhoukoudian *Homo erectus* determined with $^{26}Al/^{10}Be$ burial dating. *Nature* 458 (2009): 198–200.

Stringer, Christopher B., Francis C. Howell, and Johann K. Melentis. The signif-icance of the fossil hominid skull from Petralona, Greece. *Jour. Arch. Sci.* 6 (1979): 235–53.

Suzuki, Hisashi, and Fuyuji Takai, eds. *The Amud Man and his Cave Site.* Tokyo: University of Tokyo Press, 1970.

Swisher, Carl C. III, et al. Age of the earliest known hominids in Java, Indonesia. *Science* 263 (1994): 1118–21.

Swisher, Carl C. III, et al. Latest *Homo erectus* of Java: potential contemporaneity with *Homo sapiens* in Southeast Asia. *Science* 274 (1996): 1870–74.

Tattersall, Ian, and Niles Eldredge. Fact, theory and fantasy in human paleontology. *Amer. Scientist* 65 (1977): 204–11.

CHAPTER 8: TURKANA, THE AFAR, AND DMANISI

Research on the Nariokotome skeleton is summarized in Walker and Leakey (1993). A different viewpoint is represented by Graves et al. (2010). Wrangham (2009) ar-gues for early cooking. Early handaxes were reported by Lepre et al. (2011). The Olorgesailie hominid was reported by Potts et al. (2004); coexisting East Turkana lineages by Spoor et al. (2007); and Black Skull by Walker and Leakey (1988). New Hadar *Australopithecus afarensis* material was monographed by Kimbel, Rak, and Johanson (2004); new palate by Kimbel, Johanson, and Rak (1997). Date for Bodo given by Clark et al. (1994). Maka finds reported by White et al. (1984, 2000), and Aramis *Ardipithecus ramidus* by White, Suwa, and Asfaw (1994). The skeleton of the latter was exhaustively described by the Middle Awash Group (2009). *Ardipithecus kadabba* was described by Haile-Selassie, Suwa, and White (2004). *Ororin tugenensis* was described by Senut et al. (2001), *Sahelanthropus tchadensis* by Brunet et al. (2002), and *Australopithecus garhi* by Asfaw et al. (1999). Bouri cutmarked bones were reported by Heinzelin et al. (1999), Daka skullcap by Asfaw et al. (2002), Herto crania by White et al. (2003), and Herto context by Clark et al. (2003). The Dikika infant was announced by Alemseged et al. (2005), and cutmarked bone by McPherron et al. (2010). For stable isotope analyses, see Sponheimer and Lee-Thorp (2007). *Australopithecus anamensis* was described by Leakey et al. (1995), and amplified by Leakey et al. (1998). Kimbel et al. (2006) suggested continuity between *Australopithecus anamensis* and *Au. afarensis*. *Kenyanthropus platyops* was named by Leakey et al. (2001). The first Dmanisi jaw was described by Gabunia and Vekua (1995); the second quote is from Vekua et al. (2002); the third quote is from the description of the most recent cranium by Lordkipanidze et al. (2013); postcranials were described by Lordkipanidze et al. (2007). Review by de Lumley and Lordkipanidze (2006). Wood and Collard (1999) reconsidered the genus *Homo*, and Collard and Wood (2014) reject the inclusion of the new Dmanisi material in *Homo*.

Alemseged, Zeresenay, et al. A new hominin from the Basal Member of the Hadar Formation at Dikika, Ethiopia, and its geological context. *Jour. Hum. Evol.* 49 (2005): 499–514.

Asfaw, Berhane, et al. *Australopithecus garhi:* a new species of early hominid from Ethiopia. *Science* 284 (1999): 629–35.

Asfaw, Berhane, et al. Remains of *Homo erectus* from Bouri, Middle Awash, Ethiopia. *Nature* 416 (2002): 317–20.

Brunet, Michel, et al. A new hominid from the Upper Miocene of Chad, Central Africa. *Nature* 418 (2002): 145–51.

Clark, J. Desmond, et al. African *Homo erectus:* old radiometric ages and young Oldowan assemblages in the Middle Awash Valley, Ethiopia. *Science* 264 (1994): 1907–10.

Clark, J. Desmond, et al. Stratigraphic, chronological and behavioural contexts of Pleistocene *Homo sapiens* from Middle Awash, Ethiopia. *Nature* 423 (2003): 747–52.

Collard, Mark, and Bernard Wood. Defining the genus *Homo*. In *Handbook of Paleoanthropology*, vol. 3, 2nd ed. edited by W. Henke and I. Tattersall, 2107–2144. Heidelberg: Springer, 2014.

de Lumley, Marie-Antoinette, and David Lordkipanidze. L'homme de Dmanissi (*Homo georgicus*), il y a 1 810 000 ans. *Paléontologie humaine et Préhistoire* 5 (2006): 273–81.

Gabunia, Leo, and Abesalom Vekua. A Plio-Pleistocene hominid from Dmanisi, East Georgia, Caucasus. *Nature* 373 (1995): 509–14.

Graves, Ronda R., et al. Just how strapping was KNM-WT 15000? *Jour. Hum. Evol.* 59 (2010): 542–54.

Haile-Selassie, Yohannes, Gen Suwa, and Tim D. White. Late Miocene teeth from Middle Awash, Ethiopia, and early hominid dental evolution. *Science* 303 (2004): 1503–5.

Heinzelin, Jean de, et al. Environment and behavior of 2.5-million-year-old Bouri hominids. *Science* 284 (1999): 625–29.

Kimbel, William H., Donald C. Johanson, and Yoel Rak. Systematic assessment of a maxilla of *Homo* from Hadar, Ethiopia. *Amer. Jour. Phys. Anthropol.* 103 (1997): 235–62.

Kimbel, William, Yoel Rak, and Donald C. Johanson. *The skull of Australopithecus afarensis.* New York: Oxford University Press, 2004.

Kimbel, William, et al. Was *Australopithecus anamensis* ancestral to *A. afarensis?* a case of anagenesis in the hominin fossil record. *Jour. Hum. Evol.* 51 (2006): 134–52.

Leakey, Meave G., et al. New four-million-year-old hominid species from Kanapoi and Allia Bay, Kenya. *Nature* 376 (1995): 565–71.

Leakey, Meave G., et al. New specimens and confirmation of an early age for *Australopithecus anamensis. Nature* 393 (1998): 62–66.

Leakey, Meave G., et al. New hominin genus from eastern Africa shows diverse middle Pliocene lineages. *Nature* 410 (2001): 433–40.

Lepre, Christopher J., et al. An earlier origin for the Acheulean. *Nature* 447 (2011): 82–85.

Lordkipanidze, David, et al. Postcranial evidence from early *Homo* from Dmanisi, Georgia. *Nature* 449 (2007): 305–10.

Lordkipanidze, David, et al. A complete skull from Dmanisi, Georgia, and the evolutionary biology of early *Homo. Science* 342 (2013): 326–31.

McPherron, Shannon, et al. Evidence for stone-tool-assisted consumption of animal tissues before 3.39 million years ago at Dikika, Ethiopia. *Nature* 466 (2010): 857–60.

Middle Awash Group. *Ardipithecus ramidus. Science* 236 (2009): 1–106.

Potts, Richard, et al. Small Mid-Pleistocene hominin associated with East African Acheulean technology. *Science* 305 (2004): 75–78.

Senut, Brigitte, et al. First hominid from the Miocene (Lukeino Formation, Kenya). *C. R. Acad. Sci. Paris, Earth Planet. Sci.* 332 (2001): 137–44.

Sponheimer, Matt, and Julia Lee-Thorp. Hominin paleodiets: the contribution of stable isotopes. In *Handbook of Paleoanthropology,* vol. 1, edited by Winfried Henke and Ian Tattersall, 554–85. Heidelberg: Springer, 2007.

Spoor, Fred, et al. Implications of new early *Homo* fossils from Ileret, east of Lake Turkana, Kenya. *Nature* 448 (2007): 688–91.

Vekua, Abesalom, et al. A skull of Early *Homo* from Dmanisi, Georgia. *Science* 297 (2002): 85–89.

Walker, A. C., et al. 2.5-Myr. *Australopithecus boisei* from west of Lake Turkana, Kenya. *Nature* 322 (1986): 517–22.

Walker, Alan C., and Richard E. F. Leakey, eds. *The Nariokotome Homo erectus Skeleton.* Cambridge, MA: Harvard University Press, 1993.

White, Tim D., et al. New discoveries of *Australopithecus* at Maka in Ethiopia. *Science* 366 (1993): 261–65.

White, Tim D., et al. Jaws and teeth of *Australopithecus afarensis* from Maka, Middle Awash, Ethiopia. *Amer. Jour. Phys. Anthropol.* 111 (2000): 45–68.

White, Tim D., et al. Pleistocene *Homo sapiens* from Middle Awash, Ethiopia. *Nature* 423 (2003): 742–47.

White, Tim D., G. Suwa, and B. Asfaw. *Australopithecus ramidus,* a new species of early hominid from Aramis, Ethiopia. *Nature* 371 (1994): 306–12.

Wood, Bernard, and Mark Collard. The human genus. *Science* 284 (1999): 65–71.

Wrangham, Richard. *Catching Fire: How Cooking Made Us Human.* New York: Basic Books, 2009.

CHAPTER 9: MOLECULES AND MIDGETS

King and Wilson (1975) suggested that gene transcription was at the basis of most human-chimpanzee differences. Cann, Stoneking, and Wilson (1987) suggested a recent African origin for *Homo sapiens*. Tattersall (1986, 1992) urged more careful standards for species recognition in paleoanthropology. Tattersall and Schwartz (1999) suggested caution in interpretation of the Lagar Velho skeleton. For climate fluctuation, see Van Couvering (2014). For potential evolutionary effects, see Tattersall (2014). Hart and Sussman (2009) advocated a baboon/macaque model for early human social organization. *Australopithecus sediba* was described by Berger et al. (2010). The Flores hominid was named by Brown et al. (2004) and enlarged upon by Morwood et al. (2004, 2005).

Berger, Lee R., et al. *Australopithecus sediba:* a new species of *Homo*-like australopith from South Africa. *Science* 328 (2010): 195–204.

Brown, Peter, et al. A new small-bodied hominin from the Late Pleistocene of Flores, Indonesia. *Nature* 431 (2004): 1055–61.

Cann, Rebecca L., Mark Stoneking, and Allan C. Wilson. Mitochondrial DNA and human evolution. *Nature* 325 (1987): 31–36.

Hart, Donna, and Robert W. Sussman. *Man the Hunted: Primates, Predators, and Human Evolution. Expanded Edition.* Boulder, CO: Westview Press, 2009.

Morwood, Michael J., et al. Archaeology and age of a new hominin from Flores in eastern Indonesia. *Nature* 431 (2004): 1087–91.

Morwood, Michael J., et al. Further evidence for small-bodied hominins from the Late Pleistocene of Flores, Indonesia. *Nature* 437 (2005): 1012–17.

Tattersall, Ian. Species recognition in human paleontology. *Jour. Hum. Evol.* 15 (1986): 165–75.

Tattersall, Ian. Species concepts and species identification in human evolution. *Jour. Hum. Evol.* 22 (1992): 341–49.

Tattersall, Ian. If I had a hammer. *Scientific American,* 311, no. 3 (2014): 55–59.

Tattersall, I., and Jeffrey H. Schwartz. Hominids and hybrids: the place of Neanderthals in human evolution. *Proc. Natl. Acad. Sci. USA* 96 (1999): 7117–19.

Van Couvering, John A. Quaternary geology and environment. In *Handbook of Paleoanthropology,* vol. 1, 2nd ed., edited by W. Henke and I. Tattersall, 537–56. Heidelberg: Springer, 2014.

CHAPTER 10: NEANDERTHALS, DNA, AND CREATIVITY

See Pääbo (2014) and Disotell (2014) for an overview of the work on Neanderthal DNA. The first extraction of DNA from a Neanderthal was announced by Krings et al. (1997). Neanderthal mitochondrial genetic diversity was reviewed by Krings et al. (2000). The draft Vindija nuclear genome was published by Green et al. (2010). Neanderthal FOXP2 was characterized by Krause et al. (2007), and pigmentation alleles by Lalueza-Fox et al. (2007). Currat and Excoffier (2004, 2011) have argued against human admixture with Neanderthals as moderns spread out of Africa, but the Pääbo group has argued for it, at least at a low level (e.g., Prüfer et al., 2014; Sankararaman et al., 2014), as have Vernot and Akey (2014). The issue of methylation is addressed by Gokhman et al. (2014). Denisovan mtDNA was reported by Krause et al. (2010), and the nuclear genome by Reich et al. (2010). Possible introgression of Denisovan DNA into modern Tibetans was reported by Huerta-Sanchez et al.

(2014). The bone from the Sima de los Huesos was sequenced by Meyer et al. (2013). *Homo antecessor* was named by Bermudez de Castro et al. (1997). Happisburgh footprints were reported by Ashton et al. (2014). Carbonell et al. (2010) discuss "cultural cannibalism," and Lalueza-Fox et al. discuss the El Sidrón Neanderthals. Neanderthal reconstruction was by Sawyer and Maley (2005). Quote is from Bar-Yosef (2004). Spy calculus microfossils were reported by Henry et al. (2010). New Neanderthal dates by Higham et al. (2014). Omo 1 redating was by McDougall et al. (2005). Frontal cortex metabolic activity is discussed by Fu et al. (2011). Early beads reported by Bouzouggar et al. (2007) and d'Errico et al. (2009, 2010); Blombos plaque by Henshilwood et al. (2002); and Pinnacle Point fire treatment by Brown et al. (2009). Blombos-like motif at Diepkloof reported by Texier et al. (2010). See Tattersall (2012) for a much fuller treatment of subjects skated over here.

Ashton, Nick, et al. Hominin footprints from Early Pleistocene deposits at Happisburgh, UK. *PLoS ONE* 9, no. 2 (2014): e88329. doi:10.1371/journal.pone.0088329.

Bar-Yosef, Ofer. Eat what is there: hunting and gathering in the world of Neanderthals and their neighbors. *Int. Jour. Osteoarchaeol.* 14 (2004): 333–42.

Bermudez de Castro, Jose-Maria B., et al. A hominid from the Lower Pleistocene of Atapuerca, Spain: possible ancestor to Neandertals and modern humans. *Science* 276 (1997): 1392–95.

Bouzouggar, Abdeljalil, et al. 82,000-year-old shell beads from North Africa and implications for the origins of modern human behavior. *Proc. Natl. Acad. Sci. USA* 104 (2007): 9964–69.

Brown, Kyle S., et al. Fire as an engineering tool of early modern humans. *Science* 325 (2009): 859–62.

Carbonell, Eudald, et al. Cultural cannibalism as a paleoeconomic system in the European lower Pleistocene. *Curr. Anthropol.* 51 (2010): 539–49.

Currat, Mathias, and Laurent Excoffier. Modern humans did not admix with Neanderthals during their range expansion into Europe. *PLoS Biol.* 2 (2004): e421. doi: 10.1371/journal.pbio.0020421.g001.

Currat, Mathias, and Laurent Excoffier. Strong reproductive isolation between humans and Neanderthals inferred from observed patterns of introgression. *Proc. Natl. Acad. Sci. USA* 108 (2011): 15129–34.

d'Errico, Francesco, et al. Additional evidence on the use of personal ornaments in the Middle Paleolithic of North Africa. *Proc. Natl. Acad. Sci. USA* 106 (2009): 16051–56.

d'Errico, Francesco., et al. Pigments from Middle Paleolithic levels of es-Skhūl (Mount Carmel, Israel). *Jour. Archaeol. Sci.* 37 (2010): 3099–110.

Disotell, Todd. Biomolecules. In *Handbook of Paleoanthropology*, 2nd. ed., vol. 3, edited by W. Henke and I. Tattersall, 2015–42. Heidelberg: Springer, 2014.

Fu, Xing, et al. Rapid metabolic evolution in human prefrontal cortex. *Proc. Natl. Acad. Sci. USA* 108 (2011): 6181–86.

Gokhman, David, et al. Reconstructing the DNA methylation maps of the Neandertal and the Denisovan. *Science* 344 (2014): 523–27.

Green, Richard E., et al. A draft sequence of the Neanderthal genome. *Science* 328 (2010): 710–22.

Henry, Amanda G., Alison S. Brooks, and Dolores R. Piperno. Microfossils in calculus demonstrate consumption of plants and cooked foods in Neanderthal diets (Shanidar III, Iraq; Spy I and II, Belgium). *Proc. Natl. Acad. Sci. USA* 108 (2010): 486–91.

Henshilwood, Christopher S., et al. Emergence of modern human behavior: Middle Stone Age engravings from South Africa. *Science* 295 (2002): 1278–80.

Higham, Tom, et al. The timing and spatiotemporal patterning of Neanderthal disappearance. *Nature* 512 (2014): 306–309.

Huerta-Sanchez, Emilia, et al. Altitude adaptation in Tibetans caused by introgression of Denisovan-like DNA. *Nature* 512(2014): 194–97. doi:10.1038/nature13408.

Krause, Johannes, et al. The derived FOXP2 variant of modern humans was shared with Neandertals. *Curr. Biol.* 17, no. 21 (2007): 1908–12.

Krause, Johannes, et al. (2010). The complete mitochondrial DNA genome of an unknown hominin from southern Siberia. *Nature* 464 (2010): 894–97.

Krings, Matthias, et al. Neanderthal DNA sequences and the origin of modern humans. *Cell* 90 (1997): 19–30.

Krings, Matthias, et al. A view of Neanderthal genetic diversity. *Nature Genet.* 26 (2000): 144–46.

Lalueza-Fox, Carles, et al. A melanocortin receptor allele suggests varying pigmentation among Neanderthals. *Science* 318 (2007): 1453–55.

Lalueza-Fox, Carles, et al. Genetic evidence for patrilocal mating behavior among Neandertal groups. *Proc. Natl. Acad. Sci. USA* (2010): doi/10.1073/pnas.1011533108.

McDougall, Ian, Francis H. Brown, and John. G. Fleagle. Stratigraphic placement and age of modern humans from Kibish, Ethiopia. *Nature* 433 (2005): 733–36.

Meyer, Matthias, et al. A mitochondrial genome sequence of a hominin from Sima de los Huesos. *Nature* 505 (2013): 403–6.

Pääbo, Svante. *Neanderthal Man: In Search of Lost Genomes.* New York: Basic Books, 2014.

Prüfer, Kay, et al. The complete genome sequence of a Neanderthal from the Altai Mountains. *Nature* 505 (2013): 43–49.

Reich, David., et al. Genetic history of an archaic hominin group from Denisova Cave in Siberia, *Nature* 468 (2010): 1053–60.

Sankararaman, Sriram, et al. The genomic landscape of Neanderthal ancestry in present-day humans. *Nature* 507 (2014): 354–59.

Sawyer, Gary J., and Blaine Maley. Neanderthal reconstructed. *Anat. Rec. (New Anat.)* 283B (2005): 23–31.

Tattersall, Ian. *Masters of the Planet: The Search for Our Human Origins.* New York: Palgrave Macmillan, 2012.

Texier, Philippe J., et al. A Howiesons Poort tradition of engraving ostrich eggshell containers dated to 60,000 years ago at Diepkloof Rock Shelter, South Africa. *Proc. Natl. Acad. Sci. USA* 107 (2010): 6180–85.

Vernot, Benjamin, and Joshua M. Akey. Resurrecting surviving Neanderthal lineages from modern human genomes. *Science* 343 (2014): 1017–21.

INDEX